Field Guide to
Diffractive Optics

Yakov G. Soskind

SPIE Field Guides
Volume FG21

John E. Greivenkamp, Series Editor

SPIE PRESS
Bellingham, Washington USA

Library of Congress Cataloging-in-Publication Data

Soskind, Yakov.
 Field guide to diffractive optics / Yakov Soskind.
   p. cm. -- (The field guide series ; 21)
 Includes bibliographical references and index.
 ISBN 978-0-8194-8690-5
 1. Diffraction. 2. Optics. I. Title.
 QC415.S67 2011
 535'.42--dc23
                                                    2011018209

Published by

SPIE
P.O. Box 10
Bellingham, Washington 98227-0010 USA
Phone: +1.360. 676.3290
Fax: +1.360.647.1445
Email: books@spie.org
Web: http://spie.org

Copyright © 2011 Society of Photo-Optical Instrumentation Engineers (SPIE)

All rights reserved. No part of this publication may be reproduced or distributed in any form or by any means without written permission of the publisher.

The content of this book reflects the work and thought of the author. Every effort has been made to publish reliable and accurate information herein, but the publisher is not responsible for the validity of the information or for any outcomes resulting from reliance thereon. For the latest updates about this title, please visit the book's page on our website.

Printed in the United States of America.
First printing

Introduction to the Series

Welcome to the *SPIE Field Guides*—a series of publications written directly for the practicing engineer or scientist. Many textbooks and professional reference books cover optical principles and techniques in depth. The aim of the *SPIE Field Guides* is to distill this information, providing readers with a handy desk or briefcase reference that provides basic, essential information about optical principles, techniques, or phenomena, including definitions and descriptions, key equations, illustrations, application examples, design considerations, and additional resources. A significant effort will be made to provide a consistent notation and style between volumes in the series.

Each *SPIE Field Guide* addresses a major field of optical science and technology. The concept of these *Field Guides* is a format-intensive presentation based on figures and equations supplemented by concise explanations. In most cases, this modular approach places a single topic on a page, and provides full coverage of that topic on that page. Highlights, insights, and rules of thumb are displayed in sidebars to the main text. The appendices at the end of each *Field Guide* provide additional information such as related material outside the main scope of the volume, key mathematical relationships, and alternative methods. While complete in their coverage, the concise presentation may not be appropriate for those new to the field.

The *SPIE Field Guides* are intended to be living documents. The modular page-based presentation format allows them to be easily updated and expanded. We are interested in your suggestions for new *Field Guide* topics as well as what material should be added to an individual volume to make these *Field Guides* more useful to you. Please contact us at **fieldguides@SPIE.org**.

John E. Greivenkamp, *Series Editor*
Optical Sciences Center
The University of Arizona

Field Guide to Diffractive Optics

The Field Guide Series

Keep information at your fingertips with all of the titles in the Field Guide Series:

Field Guide to

Adaptive Optics, Tyson & Frazier

Atmospheric Optics, Andrews

Binoculars and Scopes, Yoder, Jr. & Vukobratovich

Diffractive Optics, Soskind

Geometrical Optics, Greivenkamp

Illumination, Arecchi, Messadi, & Koshel

Infrared Systems, Detectors, and FPAs, Second Edition, Daniels

Interferometric Optical Testing, Goodwin & Wyant

Laser Fiber Technology, Paschotta

Laser Pulse Generation, Paschotta

Lasers, Paschotta

Microscopy, Tkaczyk

Optical Fabrication, Williamson

Optical Lithography, Mack

Optical Thin Films, Willey

Polarization, Collett

Special Functions for Engineers, Andrews

Spectroscopy, Ball

Visual and Ophthalmic Optics, Schwiegerling

Field Guide to Diffractive Optics

Field Guide to Diffractive Optics

Recent advancements in microfabrication technologies as well as the development of powerful simulation tools have led to a significant expansion of diffractive optics and the commercial availability of cost-effective diffractive optical components. Instrument developers can choose from a broad range of diffractive optical elements to complement refractive and reflective components in achieving a desired control of the optical field.

Material required for understanding the diffractive phenomenon is widely dispersed throughout numerous literature sources. This *Field Guide* offers scientists and engineers a comprehensive reference in the field of diffractive optics. College students and photonics enthusiasts will broaden their knowledge and understanding of diffractive optics phenomena.

The primary objectives of this *Field Guide* are to familiarize the reader with operational principles and established terminology in the field of diffractive optics, as well as to provide a comprehensive overview of the main types of diffractive optics components. An emphasis is placed on the qualitative explanation of the diffraction phenomenon by the use of field distributions and graphs, providing the basis for understanding the fundamental relations and the important trends.

I would like to thank SPIE Press Manager Timothy Lamkins and Series Editor John Greivenkamp for the opportunity to write a Field Guide for one of the most fundamental physical optics phenomena, as well as SPIE Press Senior Editor Dara Burrows for her help.

My endless gratitude goes to my family: to my wife Eleanora, who had to bear additional duties during my work on this book, as well as to my children, Rose and Michael, who learned the material while helping with proofreading the manuscript.

Yakov G. Soskind
August 2011

Table of Contents

Glossary of Symbols	xi
Diffraction Fundamentals	1
The Diffraction Phenomenon	1
Scalar Diffraction	2
Paraxial Approximation	3
Fresnel Diffraction	4
Fresnel Diffraction	4
Apertures with Integer Number of Fresnel Zones	5
Fresnel Zone Plates	6
Fresnel Zone Plate Properties	7
Fresnel Phase Plates	8
Comparing Fresnel Plates and Ideal Lenses	9
Efficiency of Fresnel Plates and Ideal Lenses	10
Talbot Effect	11
Fractional Talbot Distributions	12
Fraunhofer Diffraction	13
Fraunhofer Diffraction	13
Diffraction of Waves with Finite Sizes	14
Diffraction on Ring-Shaped Apertures	15
Energy Redistribution within Diffraction Rings	16
Diffraction on Noncircular Apertures	17
Rectangular and Diamond-Shaped Apertures	18
Apodized Apertures	19
Apodized Apertures	19
Apodized Apertures with Central Obscuration	20
Field Obstruction by an Opaque Semiplane	21
Apodization with Serrated Edges	22
Serrated Apertures as Apodizers	23
Diffraction by Multiple Apertures	24
Diffraction by Multiple Apertures	24
Effects of Aperture Spacing	25
Aperture Fill Factor	26
Aperiodically Spaced Apertures	27
Resolution Limit in Optical Instruments	28
Resolution Limit in Optical Instruments	28
Superresolution Phenomenon	29

Field Guide to Diffractive Optics

Table of Contents

Superresolution with Two-Zone Phase Masks　　30
Point Spread Function Engineering　　31
Adjusting Diffraction-Ring Intensity　　32
Amplitude and Phase Filter Comparison　　33
Vortex Phase Masks　　34
Combining Amplitude and Vortex Phase Masks　　35

Diffractive Components　　**36**
Diffraction Gratings　　36
Volume Bragg Gratings　　37
Polarization Dependency of Volume Bragg Gratings　　38
One-Dimensional Surface-Relief Gratings　　39
GRISM Elements　　40
Two-Dimensional Diffractive Structures　　41
Holographic Diffusers　　42
Multispot Beam Generators　　43
Design of Fan-Out Elements　　44
Diffractive Beam-Shaping Components　　45
Digital Diffractive Optics　　46
Three-Dimensional Diffractive Structures　　47

Grating Properties　　**48**
Grating Equation　　48
Grating Properties　　49
Free Spectral Range and Resolution　　50
Grating Anomalies　　51
Polarization Dependency of Grating Anomalies　　52
Gratings as Angular Switches　　53
Gratings as Optical Filters　　54
Gratings as Polarizing Components　　55

Blazing Condition　　**56**
Blazing Condition　　56
Blazed Angle Calculation　　57
Optimum Blazed Profile Height　　58

Scalar Diffraction Theory of a Grating　　**60**
Scalar Diffraction Theory of a Grating　　60
Diffraction Efficiency　　61
Blaze Profile Approximation　　62

Field Guide to Diffractive Optics

Table of Contents

Extended Scalar Diffraction Theory	**63**
Extended Scalar Diffraction Theory	63
Duty Cycle and Ghost Orders	64
Extended Scalar versus Rigorous Analysis	65
Gratings with Subwavelength Structures	**66**
Gratings with Subwavelength Structures	66
Blazed Binary Gratings	67
Relative Feature Size in the Resonant Domain	68
Effective Medium Theory	69
Scalar Diffraction Limitations and Rigorous Theory	70
Rigorous Analysis of Transmission Gratings	**71**
Analysis of Blazed Transmission Gratings	71
Polarization Dependency and Peak Efficiencies	72
Peak Efficiency of Blazed Profiles	73
Wavelength Dependency of Efficiency	74
Efficiency Changes with Incident Angle	75
Diffraction Efficiency for Small Feature Sizes	76
Polychromatic Diffraction Efficiency	**77**
Polychromatic Diffraction Efficiency	77
Monolithic Grating Doublet	78
Spaced Grating Doublet	79
Monolithic Grating Doublet with Two Profiles	80
Diffractive and Refractive Doublets: Comparison	81
Efficiency of Spaced Grating Doublets	**82**
Efficiency of Spaced Grating Doublets	82
Sensitivity to Fabrication Errors	83
Facet Width and Polarization Dependency	84
Sensitivity to Axial Components Spacing	85
Frequency Comb Formation	86
Diffractive Components with Axial Symmetry	**87**
Diffractive Components with Axial Symmetry	87
Diffractive Lens Surfaces	88
Diffractive Kinoforms	89
Binary Diffractive Lenses	90
Optical Power of a Diffractive Lens Surface	91
Diffractive Surfaces as Phase Elements	92
Stepped Diffractive Surfaces	93
Properties of Stepped Diffractive Surfaces	94

Field Guide to Diffractive Optics

Table of Contents

Multi-order Diffractive Lenses	95
Diffractive Lens Doublets	96
Diffractive Surfaces in Optical Systems	**97**
Diffractive Lens Surfaces in Optical Systems	97
Achromatic Hybrid Structures	98
Opto-thermal Properties of Optical Components	99
Athermalization with Diffractive Components	100
Athermalization with SDSs	101
Appendix: Diffractive Raytrace	**102**
Equation Summary	**105**
Bibliography	**111**
Index	**114**

Glossary of Symbols and Acronyms

ASMA	aperiodically spaced multiple apertures
B	base length of a PRISM
C_R	incident wave obliquity factor
C_S	diffracted wave obliquity factor
CGH	computer-generated hologram
d	diameter of central obscuration
d_B	Bragg plane spacing
d_g	grating period or groove spacing
d_i	step width of ith zone
D	aperture diameter or lateral size
D_A	Airy disk diameter
D_n	material dispersion
D_0	lens clear aperture diameter
DC	duty cycle
DDO	digital diffractive optics
DLS	diffractive lens surface
e	aperture obscuration
$E(\rho, \varphi)$	complex electric field in polar coordinates
$\boldsymbol{E}_\perp$	electric field normal to the grating grooves
$\boldsymbol{E}_\parallel$	electric field parallel to the grating grooves
f	focal length of a lens
f_0^D	nominal focal length of a diffractive surface
FDTD	finite difference time domain
FWHM	full width at half maximum
FPP	Fresnel phase plate
FZP	Fresnel zone plate
h	grating profile depth or height
h_i	step height of ith zone
h_m	profile height of a multi-order diffractive lens
h_{opt}	optimum grating profile depth (height)
h_{SDS}	step height of SDS
HOE	holographic optical element
$\vec{i}$	unit vector codirectional with x axis
$I(r)$	radial intensity distribution
IR	infrared
$\vec{j}$	unit vector codirectional with y axis
$J_0(\rho)$	Bessel function of the first kind of the zero order
$J_1(\rho)$	Bessel function of the first kind of the first order

Field Guide to Diffractive Optics

Glossary of Symbols and Acronyms

$\vec{k}$	unit vector codirectional with z axis
$k(x,y)$	wave vector of the propagating wavefront
k_0	wavenumber
L	observation distance
L_T^m	Talbot distance of order m
LED	light-emitting diode
m	diffraction order
n	refractive index of optical material
n_1	refractive index before optical interface
n_2	refractive index after optical interface
n_d	refractive index of diffraction grating layer
n_p	refractive index of prism material
$n_\perp$	effective index for the electric field $\boldsymbol{E}_\perp$
$n_\parallel$	effective index for the electric field $\boldsymbol{E}_\parallel$
N	number of binary levels
N_F	Fresnel zone number
N_g	number of grating (groove) facets
N_k	number of kinoform zones
OPD	optical path difference
PSF	point spread function
$\vec{q}$	vector orthogonal to the grating plane of symmetry at the point of intersection
Q	grating "thickness" parameter
$\vec{r}$	vector normal to the grating surface at the incoming ray intersection point
$\vec{r}_{12}$	vector connecting points in two lateral planes
R_0	substrate radius of curvature
$\vec{S}$	propagation direction vector before diffractive surface
$\vec{S}'$	propagation direction vector after diffractive surface
SDS	stepped diffractive surface
SPDT	single point diamond turning
t	axial spacing of a grating doublet
T	effective thickness of volume phase grating
t_b	thickness of a binary lens level
T_g	geometrical transmission pattern of a facet
TE	transverse electric

Field Guide to Diffractive Optics

Glossary of Symbols and Acronyms

TM	transverse magnetic
$U(x,y,z)$	complex field amplitude
UV	ultraviolet
VBG	volume Bragg grating
VLSI	very large-scale integration
$W(x,y)$	propagating wavefront
W_g	grating width
$\vec{z}$	vector normal to the two reference planes
α	coefficient of thermal expansion
α_B	angle of the incident light after refraction into the volume phase medium
α_d	deflection angle
β	diffracted angle inside the volume phase medium
γ	angle between the Bragg planes and the incident light
δ	minimum feature size of the diffractive component
ϵ	grating minor (secondary) facet angle
ζ	fill factor of radiation
η	normalized diffraction efficiency
η_m	diffraction efficiency in mth diffraction order
η_M	diffraction efficiency of a grating with M facets
η_P	diffraction efficiency of P-polarized light
η_S	diffraction efficiency of S-polarized light
θ_d	diffraction angle
θ_i	angle of incidence
θ_m	mth order diffraction angle
θ_m^L	mth order diffraction angle in Littrow mount
$\theta_m^{\lambda_l}$	mth order diffraction angle of the wavelength λ_l
$\theta_m^{\lambda_s}$	mth order diffraction angle of the wavelength λ_s
θ_φ	incidence angle with respect to the grating facet
λ	wavelength of light
λ_b	blazing wavelength
λ_b^L	blazing wavelength in Littrow configuration (mount)
λ_s	the shortest wavelength within the spectral range
λ_l	the longest wavelength within the spectral range
λ_0	design wavelength

Field Guide to Diffractive Optics

Glossary of Symbols and Acronyms

ξ	opto-thermal coefficient of a surface
ρ	radial coordinate
υ	volume phase grating parameter
φ	grating primary facet angle
φ_b	grating facet blaze angle
φ_i	facet angle of ith kinoform zone
φ_p	grating passive facet angle
φ_{max}	maximum value of facet blazed angle
ϕ	optical path difference
Φ^{AH}	optical power of an achromatic hybrid
Φ^D	optical power of a diffractive surface or lens
Φ^H	optical power of a hybrid surface or lens
Φ^R	optical power of a refractive surface or lens
Φ^{SDS}	optical power of SDS
Φ^{SDS}_{eff}	effective optical power of SDS
$\psi(r)$	radial phase profile of a diffractive surface
$\psi(x,y)$	phase profile of a diffractive surface
Δf_{chr}	axial or longitudinal chromatic aberration
ΔH_{chr}	lateral or transverse chromatic aberration
$\Delta \lambda$	spectral bandwidth
$\Delta \lambda_{FSR}$	free spectral range of a grating
Δn	refractive index modulation
Λ	grating parameter

Field Guide to Diffractive Optics

The Diffraction Phenomenon

Diffraction is a fundamental wave phenomenon that explains numerous spatial radiation propagation effects that cannot be explained by geometrical optics, including the "bending" of light that leads to light presence in geometrical shade behind opaque objects. It also describes the propagation of waves with a finite spatial extent.

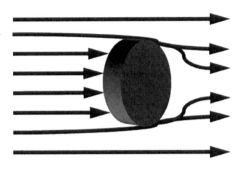

Diffraction imposes fundamental limits to the **resolution** and to the power density at the **focus** of an optical system. The diffraction phenomenon occurs over a broad range of the electromagnetic spectrum, including ultraviolet (UV), visible, infrared (IR), and radio waves. Therefore, in this field guide the terms "light" and "radiation" will be used interchangeably.

The term "diffraction" originates from the Latin word *diffringere* meaning "to break into pieces" and refers to wave fragmentation after interaction with objects. The diffraction phenomenon is often manifested by intensity ripples in the propagating field due to the coherent superposition of the diffracted and non-diffracted fractions of the propagating radiation. Diffraction has been studied by several prominent physicists, including Isaac Newton, Augustin-Jean Fresnel, Christiaan Huygens, Thomas Young, and Joseph von Fraunhofer.

Field Guide to Diffractive Optics

Scalar Diffraction

From experimental observations it is known that longer wavelengths are diffracted at larger angles, and that tighter focal spots are obtained from larger-aperture lenses. This has led to the formulation of the fundamental relation for a diffraction angle θ_d being proportional to the wavelength of light λ, and inversely proportional to the lateral dimension D of the propagating wave:

$$\theta_d \propto \lambda/D$$

Solutions to a diffraction problem consider the spatial evolution of finite-sized waves and waves whose propagation was disrupted by amplitude or phase objects. Rigorous solutions to diffraction problems satisfy **Maxwell's equations** and the appropriate boundary conditions. A simpler approach is based on the **Huygens-Fresnel principle**, which defines the foundation for scalar diffraction theory.

Scalar diffraction theory assumes that the propagating field can be treated as a scalar field. The propagation of a field described by its **complex amplitude** $U(x,y,z)$ in free space from the object plane ($z = 0$) to the observation plane is governed by the **Helmholtz equation**:

$$\nabla^2 U(x,y,z) + k_0^2 U(x,y,z) = 0$$

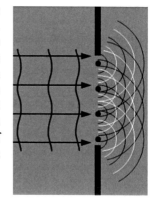

in which $k_0 = |k_0| = 2\pi/\lambda_0$ is the free space **wave number**. According to Huygens' principle, the propagating field at the aperture is considered as a superposition of several secondary point sources with spherical wavefronts. Fresnel stated that intensity distribution after the aperture is the result of interferometric interaction between the Huygens point sources.

Field Guide to Diffractive Optics

Paraxial Approximation

Monochromatic field distribution at the output plane $U(x_2, y_2)$ in accordance with the **Huygens-Fresnel principle** is calculated based on the field at the input plane $U(x_1, y_1, z_1)$ employing **Kirchhoff's diffraction integral**:

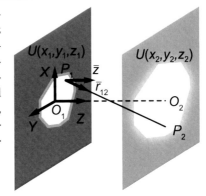

$$U(x_2, y_2, z_2) \propto \iint \frac{1 + \cos(\bar{z}, \bar{r}_{12})}{2i\lambda r_{12}} \exp(ikr_{12}) U(x_1, y_1, z_1) dx_1 dy_1$$

where $\bar{z}$ is the vector normal to the two reference planes, and $\bar{r}_{12}$ is a vector connecting the points (x_1, y_1, z_1) and (x_2, y_2, z_2) in the two planes.

The **paraxial approximation** assumes small propagation angles. In the case of the paraxial approximation, $\cos(\bar{z}, \bar{r}_{12}) \cong 1$, and $|\bar{r}_{12}| \cong z_{12}$ yield the **Fresnel diffraction integral**:

$$U(x_2, y_2, z_2) \propto \frac{\exp(ikz_{12})}{i\lambda z_{12}} \iint \exp\left\{\frac{ik}{2z_{12}}\left[(x_2-x_1)^2 + (y_2-y_1)^2\right]\right\}$$
$$\times U(x_1, y_1, z_1) dx_1 dy_1$$

The sufficient condition for the above **Fresnel approximation** is defined as

$$(z_2 - z_1)^3 \gg \pi/4\lambda \left[(x_2-x_1)^3 + (y_2-y_1)^3\right]^2_{max}$$

Large propagation distances of $z_{12} \gg \pi(x_2^2 + y_2^2)/\lambda_0$ lead to **Fraunhofer approximations** as a Fourier transform of the object field, also called **Fraunhofer diffraction** or **far-field distribution**:

$$U(x_2, y_2, z_2) \propto \frac{\exp(ikz_{12})}{i\lambda z_{12}} \exp\left[\frac{ik(x_1^2 + y_1^2)}{2z_{12}}\right] \times$$
$$\iint \exp\left[\frac{ik(x_2 x_1 + y_2 y_1)}{2z_{12}}\right] U(x_1, y_1, z_1) dx_1 dy_1$$

Field Guide to Diffractive Optics

Fresnel Diffraction

The Fresnel diffraction phenomenon is observed at finite distances from the objects that interact with the propagating field. For fields with circular symmetry, the on-axis intensity distribution depends on the number of **Fresnel zones** contained in the propagating field, as observed from the given on-axis **observation point**.

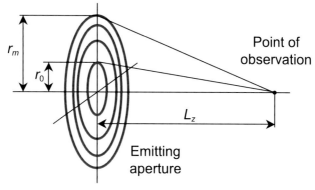

The **Fresnel zone number** N_F contained within the emitting aperture as viewed from the observation point at distance L_z is defined as a number of half waves contained in the optical path difference between the outer and inner margins of the emitting structure with respective radii r_m and r_0:

$$N_F = \frac{2}{\lambda}\left(\sqrt{L_z^2 + r_m^2} - \sqrt{L_z^2 + r_0^2}\right) \cong \frac{r_m^2 - r_0^2}{\lambda L_z}$$

The number of Fresnel zones in an unobstructed circular emitting aperture ($r_0 = 0$) with outer diameter D is

$$N_F \cong D^2/4\lambda L_z$$

The outer diameter D of a circular unobstructed emitting aperture ($r_0 = 0$) containing N_F Fresnel zones, as well as the inner diameter for a Fresnel zone with respective index ($N_F + 1$) for the observation point at a distance L_z is

$$D = \sqrt{N_F \lambda (4L_z + N_F \lambda)} \cong \sqrt{4N_F \lambda L_z}$$

Field Guide to Diffractive Optics

Apertures with Integer Number of Fresnel Zones

The axial distance from an unobstructed circular emitting aperture with the outer diameter D containing N_F Fresnel zones is found by

$$L_z = \left[D^2 - (N_F \lambda)^2\right] / 4 N_F \lambda$$

Fresnel diffraction patterns produced by emitting apertures with an odd number of zones exhibit characteristic on-axis maxima:

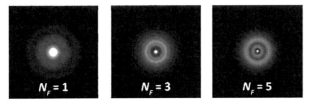

Fresnel diffraction patterns produced by emitting apertures with an even number of zones exhibit characteristic on-axis minima:

In the presence of a **central obscuration**, the distance to the observation point depends on the number of Fresnel zones contained in the emitting aperture:

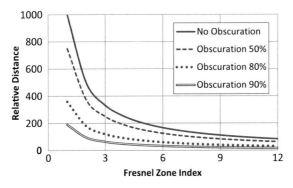

Field Guide to Diffractive Optics

Fresnel Zone Plates

A **Fresnel zone plate (FZP)** represents an amplitude mask that consists of alternating opaque and transparent rings. Each ring size corresponds to a Fresnel zone as defined by the observation point. FZPs are often employed in lieu of lenses to concentrate the propagating field into a tight on-axis spot.

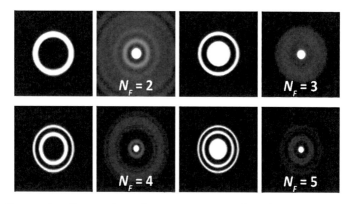

Increasing the number of zones progressively reduces the width, as well as increases the peak intensity and the total power contained in the central peak of the diffraction pattern.

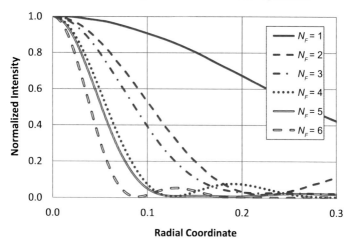

Fresnel Diffraction

Fresnel Zone Plate Properties

The increase in peak on-axis intensity for FZPs occurs for every two consecutive Fresnel numbers N_F.

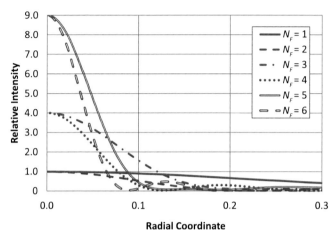

Zone plates with an equal number of transparent zones and the same aperture size produce an on-axis distribution with equal intensity and different width.

For the number of transparent FZP zones exceeding 5 and the total number of zones N_F over 11, the width value of the intensity distribution stabilizes around the **Airy distribution** width produced by an ideal lens:

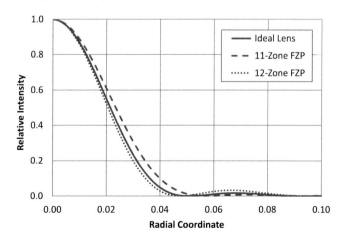

Field Guide to Diffractive Optics

Fresnel Phase Plates

A **Fresnel phase plate (FPP)** represents a phase mask that consists of transparent ring-shaped zones with alternating phase shifts by half a wavelength. Each ring size corresponds to a Fresnel zone as defined from the observation point. The zone sizes of FPPs and FZPs are identical. FPPs may be employed in lieu of lenses to concentrate the propagating field into a tight on-axis spot.

Increasing the number of zones progressively reduces the width and increases the peak intensity and the total power contained in the central peak of the pattern.

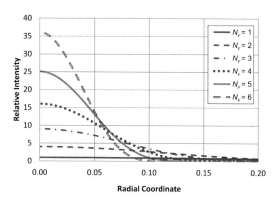

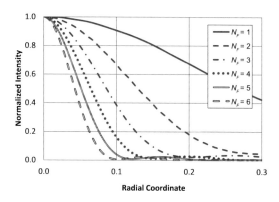

Field Guide to Diffractive Optics

Comparing Fresnel Plates and Ideal Lenses

When the number of zones in the Fresnel plates exceeds 10, the relative shape of the central peak in the normalized field distributions of the Fresnel plates and an equivalent **ideal lens** producing Airy distribution are comparable with each other, as shown for the case of Fresnel number $N_F = 12$:

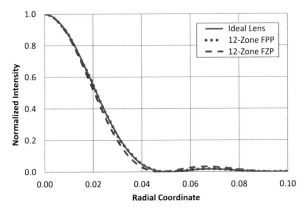

The main difference between the three distributions is in the fraction of the radiation contained outside the central peak. This is shown in logarithmic scale for the above case of $N_F = 12$ over a broader radial distance range:

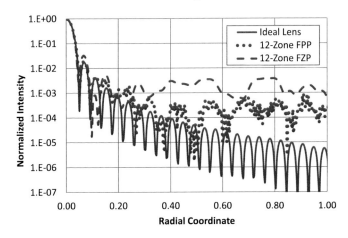

Field Guide to Diffractive Optics

Efficiency of Fresnel Plates and Ideal Lenses

FPPs concentrate significantly more energy in the central peak than do amplitude FZPs containing the same number of zones. The improvement in **diffraction efficiency** comes with increased fabrication complexity of the phase plates. At the same time, FPPs are less efficient than focusing lenses with equal apertures and F-numbers. In the case of FZPs and FPPs, the light outside the central peak is spread over a broad area:

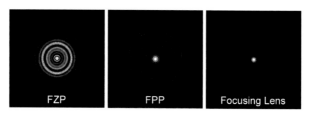

The central peaks of FZP, FPP, and an ideal lens contain 7.8%, 34.1%, and 83.8% of the total radiation power propagating through the aperture, respectively.

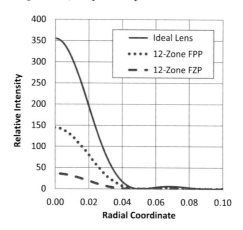

The differences between Fresnel plates and an ideal lens are summarized in the following table:

Component Type	Relative Peak Value	Center Peak Fractional Power
12-Zone FZP	0.10	7.8%
12-Zone FPP	0.41	34.1%
Equivalent Lens	1.00	83.8%

Field Guide to Diffractive Optics

Talbot Effect

The **Talbot effect**, often called Talbot imaging, occurs when a periodic array of apertures is illuminated with a coherent source. The Talbot effect is observed in the Fresnel domain and was discovered in 1836 by Henry Talbot.

Observation planes that contain **Talbot images** are called **Talbot planes**. An array of apertures with spacing d produces Talbot images of integer order m at **Talbot distances** L_T^m, defined as $L_T^m = 2md^2/\lambda$.

The figure shows the initial near-field array (left), the phase (center), and intensity distribution (right) of an image located at a Talbot distance of $L_T^1 = 2d^2/\lambda$.

Fractional Talbot distributions are produced at particular fractional Talbot distances.

Talbot half distributions with intensity maxima laterally shifted by $d/2$ from the original array are observed at the distances

$$L_T^{0.5m} = L_T^m - d^2/\lambda = 2(m - 0.5)d^2/\lambda$$

The figure below shows the initial near-field array (left), the phase (center), and the intensity distribution (right) of an image at a half Talbot distance $L_T^{0.5} = d^2/\lambda$:

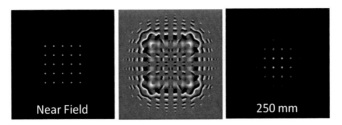

Fractional Talbot Distributions

Fractional Talbot distributions with higher spacing frequencies in the image planes can be produced. Two sets of double-frequency distributions with spacing of $d/2$ are produced at quarter distances:

$$L_T^{0.25} = L_T^{0.5} \pm d^2/2\lambda.$$

The first figure shows the phase (center graph) and intensity distribution (right graph) of a double-frequency image with $d/2$ spacing at a quarter Talbot distance $L_T^{0.25} = d^2/2\lambda$.

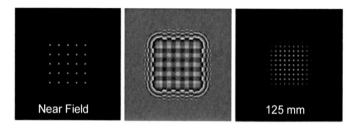

The second figure shows a quadruple-frequency distribution at the Talbot distance $L_T^{0.125} = d^2/4\lambda$ with $d/4$ intensity maxima spacing.

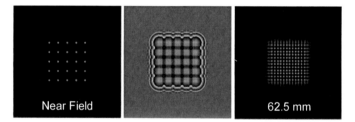

In practical applications, the finite-aperture size and spacing in the array leads to spatial degradation of the Talbot images, which is especially noticeable at the margins of the field distributions.

Field Guide to Diffractive Optics

Fraunhofer Diffraction

Far-field distribution can be conveniently observed in the vicinity of the focal plane of a lens.

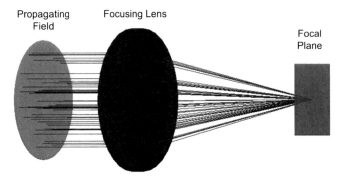

The **lens transfer function** is defined as the following phase factor:

$$U(x_1, y_1, z_1)_l = \exp\left[-i\frac{k}{2f}(x_1^2 + y_1^2)\right]$$

The field in the back focal plane of a lens is found by substituting the above phase factor into the Fresnel diffraction integral:

$$U(x_2, y_2, f) \propto \frac{\exp(ikf)}{i\lambda f} \exp\left[i\frac{k}{2f}(x_2^2 + y_2^2)\right] \times$$

$$\iint \exp\left[-\frac{ik(x_2 x_1 + y_2 y_1)}{2f}\right] U(x_1, y_1, z_1) dx_1 dy_1$$

The shape and size of the Fraunhofer diffraction patterns depend on the transfer function of a lens and are used in optical metrology for alignment purposes, as well as to identify lens imperfections. During a **star test**, an image of a source located at infinity, such as a star, produces a **point spread function** (**PSF**) in the focal plane of a lens.

The PSF size and shape from a well-corrected lens are dominated by diffraction effects, and the lens performance is called **diffraction limited**. The PSF produced by a diffraction-limited lens is called an **Airy pattern**.

Diffraction of Waves with Finite Sizes

Propagating wavefronts with finite lateral dimensions produce **Fraunhofer diffraction** patterns that are observed in the **far field** and depend on the pattern shape and size. The amplitude and phase of the field in the Fraunhofer zone are defined by a Fourier transform of the propagating field.

One of the most common Fraunhofer diffraction patterns is produced by circular apertures. The field distribution is called an **Airy pattern** after George Airy, who was the first to analytically define the intensity distribution.

A plane wave with diameter D, intensity I_0, and wavelength λ produces an Airy pattern in the focal plane of a lens with focal length f, whose radial intensity is

$$I(r) = I_0 \left(\frac{\pi D^2}{2\lambda f}\right)^2 \left[\frac{J_1\left(\frac{\pi D}{\lambda f}r\right)}{\left(\frac{\pi D}{\lambda f}r\right)}\right]^2$$

Parameter r is the radial coordinate, and the peak value of the above distribution $I(0) = I_0 \left(\pi D^2/4\lambda f\right)^2$ is found on axis. The near-field aperture distribution and the corresponding Airy pattern are shown below:

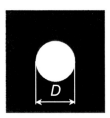
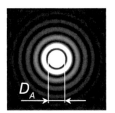

Near-field distribution　　　　　Airy pattern

The central disk of the pattern, the **Airy disk**, has a size of $D_A \cong 2.44\lambda f/D$ and contains about 84% of the total pattern energy. A fraction of the total energy contained within a circle of radius $\rho = \pi D r/\lambda f$ is calculated as

$$E(\rho) = 1 - [J_0(\rho)]^2 - [J_1(\rho)]^2$$

where $J_0(\rho)$ and $J_1(\rho)$ are Bessel functions of the first kind, of the zero and the first order, respectively.

Field Guide to Diffractive Optics

Diffraction on Ring-Shaped Apertures

Central **beam obscuration** reduces the width of the central lobe of the diffracted pattern and may be employed to improve the **resolution** of an optical system. For an aperture diameter D and the central beam obscuration diameter d, the aperture obscuration value can be defined as $e = d/D$.

The graph below shows changes in the total field power, the relative size of the central core, and the relative fraction of the total power diffracted outside the central core.

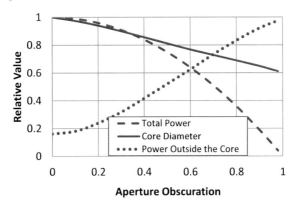

The reduction in power contained in the central core of the diffracted pattern significantly outpaces the reduction in the central core size.

For example, when the obscuration is $e = 0.7$, the fraction of the total incident energy in the diffraction rings outside the **central core** is 73.3%, and the width of the central lobe is $0.73D_A$, where D_A was defined earlier as the Airy disk diameter for an unobstructed aperture.

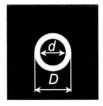
Ring-shaped near-field distribution

Circular far-field distribution

Energy Redistribution within Diffraction Rings

The Fraunhofer diffraction pattern for a circular aperture with **central obscuration** e is defined analytically as

$$I(q) = I_0 \left[(1-e^2) \frac{qD}{2} \right]^2 \left[\frac{J_1(q)}{q} - e \frac{J_1(eq)}{q} \right]^2$$

With an increase in obscuration e, the peak intensity in the first diffraction ring increases, and the maxima shift toward the center:

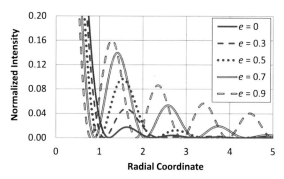

Intensity in the higher-order diffraction rings depends on the obscuration e and is not monotonic. For obscurations $e < 0.3$, the peak intensity of the second ring is reduced, while the peak intensity of the third ring is increased with an increase in obscuration:

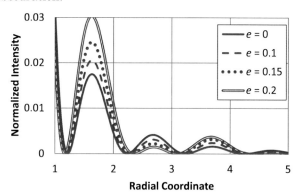

For high obscuration values ($e > 0.7$), the peak intensity in the diffraction rings is reduced with the ring order.

Diffraction on Noncircular Apertures

For near-field **elliptical distributions**, the respective Fraunhofer pattern is also elliptical, with the longer ellipse axis in the diffraction pattern orthogonal to the longer aperture axis in the near field:

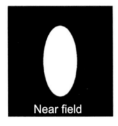

Deviation of the aperture shape from circular symmetry manifests as a distortion of the Fraunhofer pattern and may be effectively used during the inspection process.

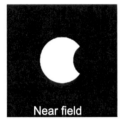
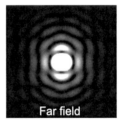

The offset of the obscuration also causes Fraunhofer **pattern distortions**.

Decentering of the obscuration is manifested by **contrast reduction** between the central core and the first diffraction ring.

Field Guide to Diffractive Optics

Rectangular and Diamond-Shaped Apertures

The lateral size of the far-field diffraction pattern is inversely proportional to the size of the near-field distribution. A diffraction pattern produced in the focal plane of a lens by a **rectangular aperture** with lateral dimensions a and b is defined as

$$U(x_2, y_2, f) = (ab)^2 \operatorname{sinc}^2\left(i\frac{kax_2}{2f}\right) \operatorname{sinc}^2\left(i\frac{kay_2}{2f}\right)$$

Rectangular aperture

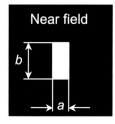
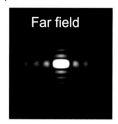

Radiation in the Fraunhofer pattern is spread in directions normal to the edges of the near-field apertures, as shown in the following graphs:

Diamond-shaped aperture

Diamond-shaped obscuration

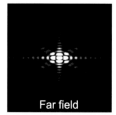

Field Guide to Diffractive Optics

Apodized Apertures

Diffraction on a finite-sized aperture causes a fraction of the propagating energy to spread outside the central core of the Fraunhofer diffraction pattern into the outside rings or nodes. In many applications, this may also lead to adverse effects such as difficulties in detecting weak signals, channel crosstalk, changes in photodetector responsivity, etc.

Apodization is an important technique employed to reduce the amount of light diffracted into the diffraction rings or nodes. **Soft-edge apertures** with the amplitude transmission function gradually changing in the vicinity of the edge from 100% transmissive to completely opaque are commonly used as apodizing structures. The following figures show a reduction in the number of diffraction rings with an increase in width of the transition edge zone of the soft-edge aperture.

Soft-edge apertures with different transition widths:

Transition 0.07D Transition 0.15D Transition 0.25D

Corresponding far-field patterns:

Far field   Far field   Far field

Apodized Apertures with Central Obscuration

Apodized **apertures with central obscuration** significantly reduce both the amount of light outside the central core and the central core diameter.

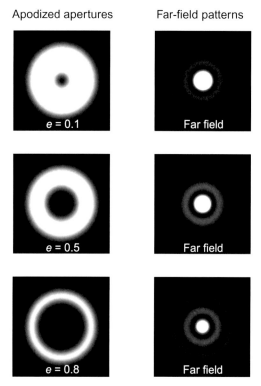

An increase in central obscuration leads to a reduction in the central core size and an increase in the relative amount of energy in the diffraction rings. For a given obscuration value e, the energy outside the central core of the soft-edge apodized apertures is significantly lower than for the apodized shapes.

Field Obstruction by an Opaque Semiplane

Diffraction of a field obstructed by an **opaque semiplane** with a straight edge occurs in numerous devices, including beam profilers, variable attenuators, and optical shutters. Replacing the sharp hard edge with a soft apodized edge significantly reduces diffractive energy spread orthogonal to the edge.

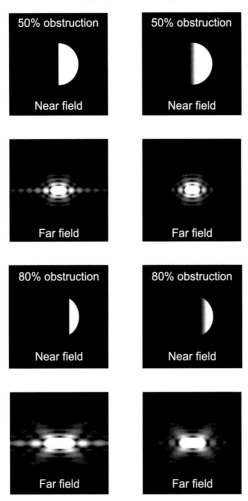

Apodizers with soft edges can significantly reduce the energy spread outside the central core into diffraction rings and nodes.

Apodization with Serrated Edges

Serrated edges can be effectively used for suppressing diffracted light in the vicinity of the central core of far-field diffraction patterns obstructed by opaque objects. Serrated edges can be employed in high-power laser applications, whereas absorbing **soft-edge apodizers** can be easily damaged under similar conditions.

The figures below show the influence of edge serration on Fraunhofer diffraction patterns for apertures with 50% obstruction. A diffraction pattern for an aperture with a straight edge (left column) is compared to the respective diffractive patterns produced by two different serrated edges.

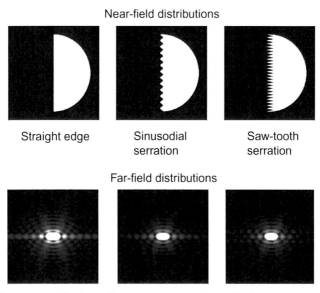

The saw-tooth-shaped serrations produce the least amount of light diffracted in the direction normal to the semi-aperture edge.

Field Guide to Diffractive Optics

Serrated Apertures as Apodizers

Serrated apertures can also be employed to suppress the diffraction of light outside the central core of the far-field diffraction pattern. This technique is especially important in high-power laser applications, when absorbing **grayscale apodizers** with soft edges can no longer be employed. Serrations may take various shapes, and radial serrations are most common.

The figures below present the field distributions for circular apertures with progressively increasing radial serration depths.

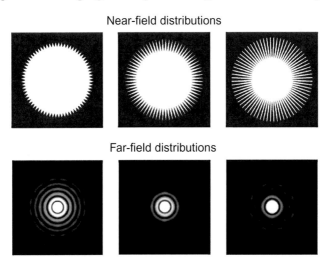

Two effects are noticeable. An increase in the serration depth leads to a significant reduction of light diffracted outside the central core. At the same time, the increase in the serration depth causes an increase in the diameter of the central core in the Fraunhofer diffraction pattern.

Field Guide to Diffractive Optics

Diffraction by Multiple Apertures

Far-field patterns produced by diffraction on **multiple apertures** depend on the size, shape, and quantity of the apertures.

Fraunhofer diffraction by multiple rectangular slits

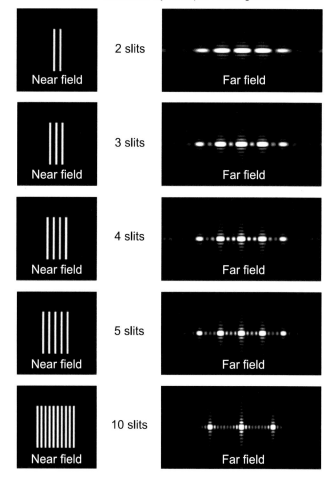

Effects of Aperture Spacing

Effects of **aperture spacing** between multiple apertures are illustrated by observing diffraction on five identical rectangular apertures with width w and height h. The aperture centers are spaced a distance d from each other.

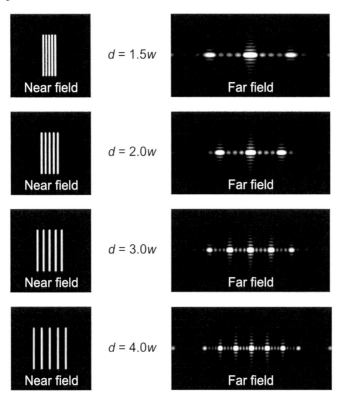

An increase in the aperture spacing d is associated with a respective increase in the number of maxima in the diffraction pattern. The spacing between the maxima and the lateral maxima widths is inversely proportional to the aperture spacing.

Aperture Fill Factor

The near-field aperture **duty cycle**, also known as a **fill factor** ζ, is defined as the ratio of the aperture width w to the aperture spacing value d ($\zeta = w/d$). The effects of the duty cycle changes on the shape of a diffraction pattern are shown for three rectangular apertures:

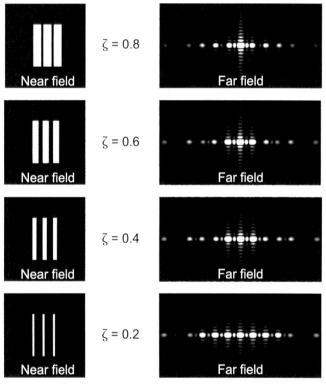

An increase in the duty cycle leads to a reduction in the number of high-peak-intensity nodes, as well as an increase in the energy concentration in the central diffraction spot.

In the limiting case of $\zeta = 1$, the pattern becomes identical to an entirely filled single rectangular aperture with a width of $3w$ and a height h.

Field Guide to Diffractive Optics

Diffraction by Multiple Apertures 27

Aperiodically Spaced Apertures

Aperiodically spaced apertures (ASAs) represent an ensemble of apertures with a variable duty cycle. ASAs are employed to suppress secondary radiation maxima in the far field.

An exemplary ASA structure consisting of 10 apertures with constant aperture width and gradually varying spacing as well as the associated far-field distribution is shown below.

Respective field distributions from a periodic structure consisting of 10 equally spaced apertures:

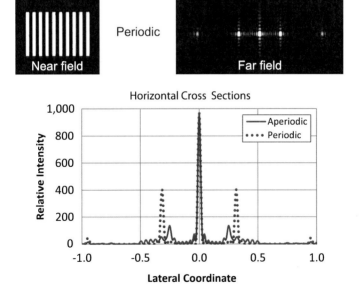

Field Guide to Diffractive Optics

Resolution Limit in Optical Instruments

An **extended object** can be considered a superposition of individual points with varying intensities across the object. For a diffraction-limited lens, each object point produces a PSF in the image plane. Intensity distribution in the image plane produced by incoherent illumination represents a **convolution** of the object intensity and the PSF.

The **Rayleigh resolution criterion** is based on distinguishing two closely spaced point objects of equal intensity when the PSF maximum of one object coincides with the first PSF minimum of the second object. The minimum separation between the two resolvable incoherent sources at the focal plane of a diffraction-limited optical system based on the Rayleigh criterion can be described as

$$d_{min}^R = 1.22 \lambda f/D$$

The **Sparrow resolution criterion** is based on an intensity minimum appearing in the joint intensity function between the two objects. It can be applied to visual observations based on the high sensitivity of the human eye to intensity differences, and in the case of two objects with equal intensity, is defined as

$$d_{min}^S = 0.95 \lambda f/D$$

In the case of **coherent illumination**, the combined field distribution is defined based on amplitude addition of the PSFs. The resolution of two objects now depends on the phase delay between the objects. For the two in-phase objects (zero phase delay) with equal intensity, the Rayleigh criterion no longer resolves the objects. The Sparrow criterion requires 1.5 times larger separation and is defined as

$$d_{min}^S = 1.46 \lambda f/D$$

For two objects with a phase delay of $\lambda/2$, the coherently added pattern has the minimum intensity value between the objects, regardless of the separation d. **Phase masks** are commonly employed to improve resolution in applications using coherent illumination, such as projection lithography.

Field Guide to Diffractive Optics

Superresolution Phenomenon

Resolution enhancements in the far field are commonly achieved using **pupil filters** or **pupil masks**. The pupil filters are located at the lens aperture stop, or at the entrance or exit pupil locations of the lens.

Lenses employing apertures with central obscuration reduce the width of the PSF central peak below the width of the central peak in the Airy pattern (also known as the **Airy disk**). PSFs with central peak width lower than the widths of the Airy disc are referred to as **superresolved PSFs**.

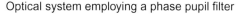
Optical system employing a phase pupil filter

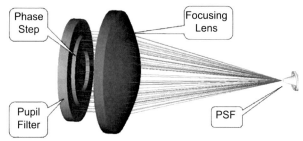

Amplitude masks, phase masks, and their combinations are employed to produce superresolved PSFs. The idea of using pupil phase masks with alternating phase delays between the neighboring ring zones to reduce the PSF width was proposed by **Toraldo di Francia** in 1952.

Reduction in the PSF central peak width is associated with a reduction in the **Strehl ratio**, which is defined as the ratio of the PSF peak value to the PSF peak value of a diffraction-limited distribution. Reduction in the PSF central peak width is also associated with diffraction of a significant fraction of the propagating energy to the outside of the central peak and an increase in the peak intensity of the diffraction rings.

Superresolution with Two-Zone Phase Masks

The simplest phase mask consists of two transparent circular zones with a relative phase delay.

PSF shapes produced by a two-zone pupil phase mask with a relative phase delay of π producing an **optical path difference (OPD)** of $\lambda/2$ are shown in the figure below for different normalized inner-zone radii R. Airy distribution corresponds to the inner-zone radius $R = 0$. The figure shows the tradeoff between the central peak width and the relative energy contained in the diffraction rings.

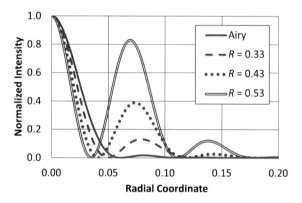

Reduction in the PSF central peak width is associated with reduction in the **Strehl ratio**, defined as the ratio of the PSF peak intensity to the peak intensity of the Airy disc.

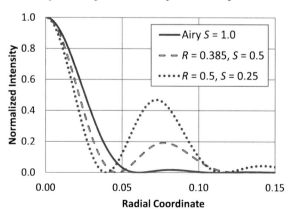

Field Guide to Diffractive Optics

Point Spread Function Engineering

The intensity in the first diffraction ring reaches the intensity of the central peak when the zone radius $R = 0.569$. When the relative zone size reaches $R \cong 0.7$, the central peak vanishes, and the PSF becomes ring shaped. For relative zone sizes over 0.84, the PSF no longer has a depression in the center of the distribution, and instead produces a flattened central peak.

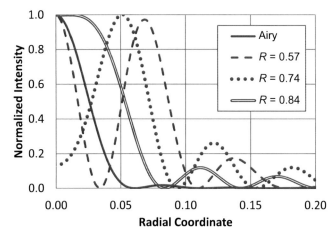

Instrumentation fields that benefit from superresolved PSFs include confocal scanning microscopy and optical data storage. In the case of scanning confocal microscopy, a desirable increase in the axial resolution is achieved by producing PSFs with reduced axial extent as compared to diffraction-limited PSFs.

Resolution-enhancement techniques constitute a subset of more general **point spread function engineering** techniques that represent an active topic in contemporary optical research. A variety of specially designed PSFs are used in both imaging and nonimaging instruments.

Vortex phase masks have been employed in solar coronagraphs, in high-resolution fluorescent depletion microscopy, and in optical tweezers for particle trapping and manipulation.

Another use of phase structures was found in **extended depth of field** imaging systems as well as for producing a special class of asymmetric **Airy beams**.

Adjusting Diffraction-Ring Intensity

It is often desirable to reduce the peak intensity in the PSF **diffraction rings**. In the case of two-zone phase masks, the reduction in core width is inevitably associated with an intensity increase of the first diffraction ring.

Increasing the number of the phase mask zones provides additional degrees of freedom in PSF design. A three-zone phase mask can significantly reduce the peak intensity in the first diffraction ring of a superresolved PSF.

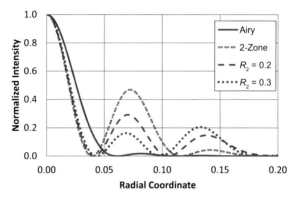

PSF engineering can significantly reduce the peak intensities in the diffraction rings. For a three-zone phase mask with the second radius $R2 = 0.886$ and Strehl ratio of 0.5, a 3× reduction in the secondary peak intensity values is achieved. The peak intensity in the diffraction rings is below 0.6% of the central peak intensity value.

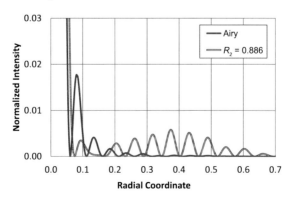

Field Guide to Diffractive Optics

Amplitude and Phase Filter Comparison

The changes in PSF shape over the inner zone radius occur more rapidly in the case of phase **pupil filters**. The following figure presents the power contained in the PSF central core as a function of the inner zone radius:

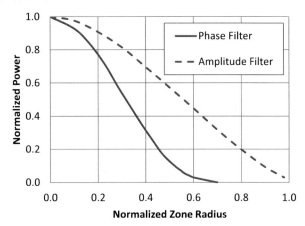

When differences in the rate of change in the PSF central core power are accounted for, the PSF widths produced using both pupil **amplitude filters** and **phase filters** result in comparable performance:

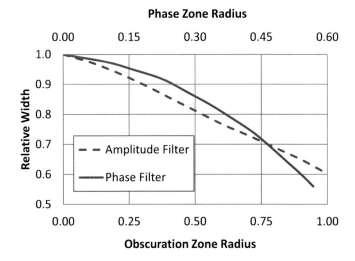

Field Guide to Diffractive Optics

Vortex Phase Masks

Vortex phase masks are extensively employed with coherent radiation to produce doughnut-shaped beams in **optical tweezers** and in **fluorescence depletion microscopy**. Vortex masks are also employed as pupil phase filters to alter the PSF of an optical system.

For an optical system with a uniformly illuminated vortex phase mask located at the pupil, the field distribution can be written as

$$E(\rho,\phi) = \begin{cases} E_0 e^{im\phi}, & \text{when } \rho \leq R_{max} \\ 0, & \text{when } \rho > R_{max} \end{cases}$$

where ρ is the radial coordinate, ϕ is the azimuthal coordinate in the transverse plane, m is the topological charge, R_{max} is the maximum pupil radius, and E_0 is the pupil amplitude.

The pupil phase profiles and the respective doughnut-shaped far-field intensity distributions for **topological charges** $m = 1$ and $m = 2$ are shown below:

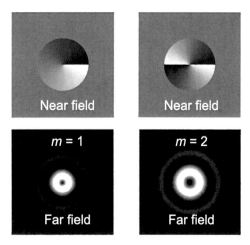

Field Guide to Diffractive Optics

Combining Amplitude and Vortex Phase Masks

The **doughnut-shaped field** size gradually increases with an increase in the topological charge of the vortex phase mask:

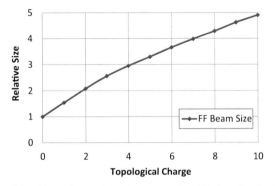

The combination of a vortex phase mask with topological charge $m = 2$ and an elliptical amplitude mask at the pupil of an optical system produces elongated **superresolved PSFs**. The figure below shows the transition from a doughnut-shaped PSF with circular amplitude mask (ellipticity $\varepsilon = 1$) to an elongated superresolved PSF ($\varepsilon = 10$):

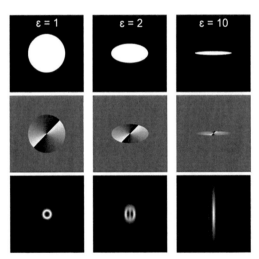

The top, central, and bottom rows correspond respectively to the amplitude masks, combinations of the two masks, and the PSFs for the ellipticities $\varepsilon = 1$, $\varepsilon = 2$, and $\varepsilon = 10$.

Field Guide to Diffractive Optics

Diffraction Gratings

Diffraction gratings are periodic diffractive structures that modify the amplitude or phase of a propagating field. **Linear gratings** represent the simplest periodic diffractive structures.

Amplitude gratings are based on the amplitude modulation of the incident wavefront and are employed in spectral regions where nonabsorbing optical materials are not available. The amplitude modulation is associated with transmission losses introduced by the grating.

Phase gratings are based on the phase modulation of the incident wavefront by introducing a periodic phase delay to the individual portions of the propagating wavefront. Phase gratings are designed to work in **transmission, reflection**, or in a bidirectional manner.

Surface-relief phase gratings are based on wavefront-division interference principles and introduce periodic phase delays to the fractions of the incident wavefront due to periodic changes of the substrate thickness.

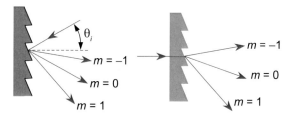

Reflective surface-relief phase grating with triangular profile

Transmissive surface-relief phase grating with triangular profile

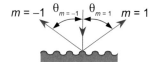

Reflective surface-relief phase grating with sinusoidal profile

Volume Bragg Gratings

Volume Bragg gratings (**VBGs**) are based on **amplitude division** interference and are designed to perform in reflection or in transmission. VBGs introduce a periodic phase delay to the propagating wavefronts associated with the periodic modulation Δn of the

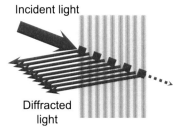

refractive index of the grating material. VBG diffraction efficiencies of **S-polarized** (**TE-polarized**) and **P-polarized** (**TM-polarized**) light are calculated as

$$\eta_S = [\sin(\upsilon)]^2$$
$$\eta_P = \{\sin[\upsilon\cos(2\gamma)]\}^2$$

where the angle 2γ is between the incident light and diffracted light inside the volume phase medium, and the parameter υ is defined as $\upsilon = \dfrac{\pi \Delta n T}{\lambda \sqrt{C_R C_S}}$.

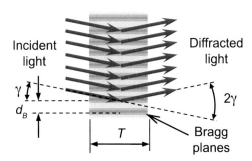

T is the effective grating thickness, C_R is the incident wave obliquity factor, and C_S is the diffracted wave obliquity factor:

$$C_R = \cos(2\gamma)$$
$$C_S = \cos(2\gamma) - \frac{\lambda}{n d_B}\tan\left(\frac{\beta - \alpha_B}{2}\right)$$

where d_B is the **Bragg plane** spacing, α_B is the incidence angle after refraction into the phase medium, and β is the angle after refraction into the volume phase medium.

Polarization Dependency of Volume Bragg Gratings

VBGs can be designed to operate either as polarization-independent or polarizing components, depending on the relation between the peak diffraction efficiency of the **S-polarization** and **P-polarization** states.

The polarization-independent operating condition is
$$\cos(2\gamma) = -\frac{2p-1}{2s-1}$$

The S-polarizing operating condition is
$$\cos(2\gamma) = -\frac{2p}{2s-1}$$

The P-polarizing operating condition is
$$\cos(2\gamma) = -\frac{2p-1}{2s}$$

The refractive index modulation Δn corresponding to the S-polarization peak is calculated as

$$\Delta n = \frac{\lambda(s-0.5)}{T}\sqrt{\cos(2\gamma)\left[\cos(2\gamma) - \frac{\lambda}{nd_B}\tan\left(\frac{\beta-\alpha_B}{2}\right)\right]}$$

The relation between the angles α_B and β is found from the grating equation
$$\sin(\alpha_B) + \sin(\beta) = \frac{\lambda}{nd_B}$$

The peak efficiency maxima for the S- and P-polarization states are satisfied when either of the following two possible conditions is valid:

$$\beta = \cos^{-1}\left(\frac{2p-1}{2s-1}\right) - \alpha_B$$

$$\beta = 180 - \cos^{-1}\left(\frac{2p-1}{2s-1}\right) - \alpha_B$$

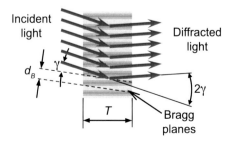

Field Guide to Diffractive Optics

One-Dimensional Surface-Relief Gratings

The evolution of diffraction grating technology has led to a variety of grating profiles and structures. The choice of a specific profile or structure is governed by fabrication costs and performance requirements.

Surface-relief gratings with **triangular grooves** are formed on top of aluminum- or gold-coated substrates by a diamond tool of a ruling engine. The grooves are defined by the

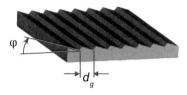

groove spacing d_g and the **facet angle** φ. Alternatively, triangular grooves are fabricated by using a grayscale mask **photoresist** exposure and a subsequent **transfer etching** into the grating substrate.

Lamellar gratings are composed of rectangular ridges of width w and height h spaced at a distance d_g from each other. They are well suited for fabrication using well-established lithographic techniques. Lamellar grating structures often have ridges that are compara-ble with the operational wave-

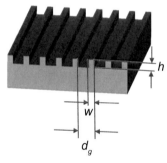

length λ and are therefore designed using rigorous diffraction techniques.

Sinusoidal grating profiles are usually etched into the substrate after being exposed to a pattern produced by two-beam interference.

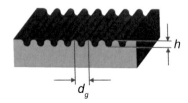

Field Guide to Diffractive Optics

GRISM Elements

The term **GRISM** refers to an integrated optical component that combines a diffraction grating and a prism in a single element. Direct-view GRISM spectrometers with constant dispersion combine the grating and the prism dispersions while providing a cancellation of the dispersion slopes.

The GRISM can be designed to satisfy the **zero-deflection** condition and to avoid deflection of the propagating radiation. The zero-deflection condition is found for the grating **blazing condition** when $\theta_i = \varphi$. The deflection an-

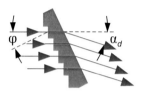

gle α_d at the exit of a planar grating with the substrate refractive index n_d is

$$\alpha_d = \sin^{-1}[n_d \sin(\varphi)] - \varphi$$

The optimum step height h_{opt} of the grating facet for the blazed wavelength λ_b in the mth diffraction order is

$$h_{opt} = \frac{m\lambda_b}{(n_d - 1)}$$

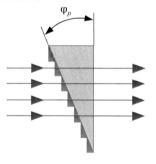

A zero-deflection GRISM is produced when the grating interface is applied to a surface of a prism with a properly defined vertex angle φ_p. The vertex angle of a prism φ_p for a zero-deflection GRISM is calculated as

$$\tan(\varphi_p) = \frac{n \sin(\varphi) - \sin(\varphi)}{\sqrt{(n_p)^2 - [n_d \sin(\varphi)]^2} - \cos(\varphi)}$$

Field Guide to Diffractive Optics

Two-Dimensional Diffractive Structures

Two-dimensional diffractive structures are composed of two-dimensional arrays of microstructures. The shape, size, and spacing may differ in the two lateral directions. Fabrication of two-dimensional gratings is performed using lithographic techniques.

Two-dimensional diffraction gratings are often made with feature sizes smaller than the operating wavelengths and are called **photonic crystals** or **artificial dielectrics**. The features may be in the form of cavities etched into the substrate or pillars elevated above the substrate level.

Rigorous diffraction techniques, such as **finite difference time domain (FDTD)**, are commonly employed to simulate the performance of photonic crystals. The two most common applications of two-dimensional gratings include antireflection surface-relief microstructures and microstructured surfaces for increased light-emitting diode (LED) light extraction and increased photovoltaic cell efficiency.

The figure shows an example of a two-dimensional diffractive structure that has antireflection properties when the feature size is less than the wavelength. An **effective refractive index** increases gradually from the index of air to the index of the substrate.

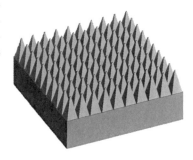

Field Guide to Diffractive Optics

Holographic Diffusers

Holographic diffusers (or **diffractive homogenizers**) are employed to control light distribution in illumination and laser systems. Diffusers represent **far-field shaping components** designed to transform coherent and incoherent radiation using multiple diffraction orders. Traditional opal glass or ground diffusers are limited to **Lambertian scatter**. Holographic diffusers allow substantial flexibility in controlling spatial illumination patterns, producing nonrotationally symmetric spatial shapes as well as desired angular distributions.

Diffusion angles can be selected within a broad range of 0.5–80 deg. Circular, oval, square, and line-shaped patterns are common. More than 90% of the incident light can be directed into the specified illumination pattern.

Holographic diffusers are cost-effectively replicated in high volume using injection or compression molding or can be embossed onto surfaces of other optical components. A master mold for a holographic diffuser is produced as a surface-relief structure by **holographic recording** with subsequent lithographic fabrication.

Illumination patterns produced by 10 LEDs

Illumination patterns produced by 10 LEDs with a holographic diffuser

Field Guide to Diffractive Optics

Multispot Beam Generators

Multispot beam generators, also known as **array beam generators** or **fan-out elements**, are diffractive elements that split an incoming laser beam into a finite number of beams with a specific intensity distribution. They belong to the class of diffractive **far-field shaping components**, or **Fourier gratings**. The field distribution in the angular space θ_x after the fan-out element is given by

$$U(\theta_x) = \sum_{n=1}^{N} A_n \exp(i\phi_n)\delta(\theta_x - \theta_n)$$

where A_n is the amplitude, ϕ_n is the phase, and θ_n is the angular coordinate of a one-dimensional spot array.

Far-field beam shapers are composed of multiple **diffractive phase cells** significantly smaller than the size of the laser beam employed during multispot beam generation. The fan-out elements are relatively insensitive to the size and position of the laser beam at the diffractive element.

Due to the small size of the phase cells, the far-field condition is satisfied starting from distances close to the multispot generator. The formed multispot pattern will scale in the far field based on the divergence angles of the generated beams.

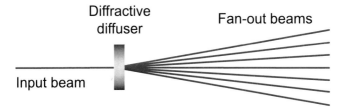

Multispot beam generators are used as image replicators, as beam splitters or beam combiners, and as spatial multiplexors in multichannel optical interconnects.

Design of Fan-Out Elements

The design of fan-out elements is based on multiple approaches, including **Dammann gratings** and multilevel binary and **kinoform** structures.

Dammann gratings represent **binary phase grating** structures. Each grating period is divided into multiple segments with phase shifts of 0 and π. The fan-out elements based on Dammann gratings are simple and require only a single mask during the fabrication process. However, the diffraction efficiency of fan-out elements based on Dammann gratings is around 80%. The figures below show the phase profile and field distribution of a three-beam fan-out element with transition points at 0.47 and 0.70.

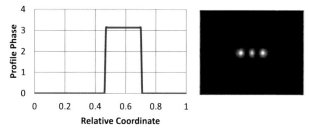

Kinoform structures with continuous profiles produce fan-out elements with the highest diffraction efficiencies. Fan-out elements designed to achieve uniformity of the resulting beam intensities at <1% have efficiency over 92%. Diffraction efficiencies of 98–99% are achieved when the intensity uniformity requirement can be relaxed. The profile shape of a kinoform fan-out element designed to split the propagating beam into nine beams is shown.

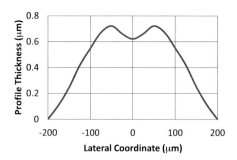

Field Guide to Diffractive Optics

Diffractive Beam-Shaping Components

Diffractive beam-shaping components are employed to modify the field distribution of coherent radiation. Diffractive **near-field shaping components** introduce spatially varying phase delay across the propagating laser beam and remap the beam onto a plane of regard using a single diffractive order. Field distributions produced with the near-field diffractive beam-shaping technique are sensitive to variations in phase and intensity of the input beam. A phase corrector placed in the plane of the remapped distribution combined with a **Fourier transform lens** is employed to extend the axial range of the shaped field along the propagation direction.

Far-field shaping components can be made as pixilated phase structures. They are referred to as **digital diffractive optical elements** and can produce intricate radiation patterns. Local intensity variations due to interference effects, known as **speckle**, are present in the field distributions generated using diffractive beam shapers. The average speckle size d at the observation plane located a distance L from the diffuser is estimated to be

$$d = L\frac{\lambda}{D}$$

where λ is the wavelength of the propagating radiation, and D is the input beam diameter. Speckle reduction in the generated pattern is achieved by employing input radiation with reduced coherence and by translating the input beam across the diffuser surface, as well as by performing spatial filtering of the pattern.

The figures below show diffractive beam shaping of a Gaussian laser beam into a square-shaped beam with uniform intensity distribution.

Field Guide to Diffractive Optics

Digital Diffractive Optics

Digital diffractive optical (DDO) elements represent periodic two-dimensional discrete phase structures that are designed using iterative procedures. Digital diffractive optical elements are also called **computer-generated holograms** and represent versatile structures, allowing for the creation of arbitrarily shaped field distributions. The phase profile of a DDO is found by solving an inverse problem wherein the input beam and the desired field distribution are used as input parameters.

Compared to **holographic optical elements**, which represent diffractive structures with continuous changes in phase profile and are recorded through interference of at least two beams, DDOs provide significantly higher flexibility in defining complex diffraction patterns.

DDOs are made by fabrication of the designed diffraction pattern onto the element substrate. The patterns are often transferred into the substrate during lithographic fabrication using a mask set. DDOs are widely used for pattern generation and beam shaping and can be cost-effectively replicated in high volumes.

The figure below shows a phase-distribution fragment of a DDO (left) employed to produce a square-shaped Fraunhofer field distribution (right).

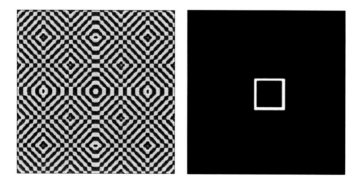

Field Guide to Diffractive Optics

Three-Dimensional Diffractive Structures

Three-dimensional diffractive structures represent the most complex microstructures that exhibit **volumetric anisotropy**.

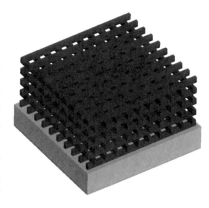

Photonic crystals and **volume Bragg gratings** (VBGs) are produced by multiple exposures to the substrate and are two examples of three-dimensional diffractive structures. In both cases, diffraction is caused by the **Bragg phenomenon**.

Three-dimensional diffractive structures present significant challenges with respect to analysis and fabrication. Simulation techniques developed for photonic crystals, such as **band diagrams**, **Bloch states**, and **Brillouin zones**, are incompatible with ray-tracing techniques commonly used in optical design. **Isofrequency diagrams**, or **wave vector diagrams**, can be employed only to define the number of propagating diffraction orders. Photonic crystal designs are concerned with confinement of the propagating field. The operating wavelength and the relative feature size of the crystal, known as the **lattice constant**, are selected below the **photonic bandgap**, ensuring an evanescent nature of all diffraction orders in reflection and in transmission.

The volumetric anisotropy of photonic crystals leads to several "unusual" propagation phenomena, such as the **superprism effect**, **supercollimation**, and **negative refraction**.

Grating Equation

The **grating equation** defines the propagation directions of radiation after interaction with the grating structure. For a plane wave incident onto the grating structure at an angle θ_i, the diffraction angle θ_m is found from the following general equation:

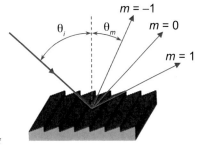

$$n_2 \sin\theta_m = n_1 \sin\theta_i + m\lambda/d_g$$

where n_1 and n_2 define the refractive indices of the material before and after the grating interface. When the diffraction order is zero ($m = 0$), the grating equation reduces to the well-known **Snell's law**. For a reflection grating located in air ($n_1 = -n_2 \cong 1$),

$$\sin\theta_m + \sin\theta_i = m\lambda/d_g$$

When the grating structure is applied to one of the surfaces of a plane-parallel plate made of refractive material, the diffraction angle at the plate exit θ_m is

$$\sin\theta_m = \sin\theta_i + m\lambda/d_g$$

Littrow mounting, also known as **autocollimation**, is a specific condition in which the incidence and diffraction angles of a reflection grating are equal ($\theta_i = \theta_m^L$). The grating equation for autocollimation is

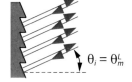

$$\sin\theta_m^L = m\lambda/2d_g$$

The grating **classical mount** corresponds to the incident plane being normal to the grating grooves. The **conical mount** occurs when the incident plane is at an angle to the grating grooves so that diffraction orders deviate from the incident plane, forming a cone.

Field Guide to Diffractive Optics

Grating Properties

The **grazing incidence** mount corresponds to steep angles of incidence required to achieve higher resolution. **Grazing mounts** are often used in **autocollimation**.

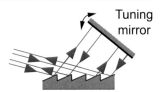

For a constant angle of incidence, the grating **angular dispersion** is inversely proportional to the cosine of the diffracted angle θ_m:

$$\frac{\partial \theta_m}{\partial \lambda} = \frac{m}{d_g n_2 \cos \theta_m}$$

The angular dispersion of a blazed reflective grating in Littrow mounting is reduced by half. The dispersion is a function of the blaze angle φ_b and is no longer dependent on the order of diffraction m:

$$\frac{\partial \theta_m^L}{\partial \lambda} = \frac{m}{2 d_g \cos \theta_m^L} = \frac{\tan(\varphi_b)}{\lambda_b^L}$$

Linear dispersion at the focal plane of a focusing objective with a focal length f is the product of the focal length and the angular dispersion of the grating.

An alternative form of grating angular dispersion is obtained using the incidence and diffraction angles:

$$\frac{\partial \theta_m}{\partial \lambda} = \frac{n_2 \sin \theta_m - n_1 \sin \theta_i}{\lambda n_2 \cos \theta_m}$$

When accounting for the **blazing condition**, the angular dispersion becomes

$$\frac{\partial \theta_m}{\partial \lambda} = \frac{1}{\left[\frac{d_g n_1 \cos(\theta_i)}{m} - \frac{\lambda_b}{\tan(\varphi_b)}\right]}$$

From the grating equation it follows that longer wavelengths are diffracted at larger angles. For the two wavelengths $\lambda_s < \lambda_l$, the relation between the respective diffraction angles is $\theta_m^{\lambda_s} < \theta_m^{\lambda_l}$.

Field Guide to Diffractive Optics

Free Spectral Range and Resolution

The free spectral range $\Delta\lambda_{FSR}$ of a grating in a given diffraction order m defines the largest bandwidth that does not overlap with the same bandwidth in adjacent orders:

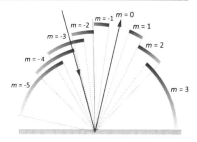

$$\Delta\lambda_{FSR} = \frac{\lambda_s}{m} = \frac{\lambda_l}{m+1}$$

For gratings operating in higher orders, the free spectral range $\Delta\lambda_{FSR}$ is significantly reduced.

Grating resolution is proportional to the product of the diffraction order m and the total number of illuminated grating grooves N_g:

$$\frac{\lambda}{d\lambda} = |m|N_g$$

The resolution of a uniformly illuminated grating is proportional to the grating width W_g:

$$\frac{\lambda}{d\lambda} = |m|\frac{W_g}{d_g} = \frac{W_g}{\lambda}|\sin\theta_m + \sin\theta_i|$$

For a given grating groove spacing d_g, the resolution can be increased either by increasing the order m, or by enlarging the grating width W_g. The upper limit of the grating resolution is defined as the number of half-wavelengths contained within the grating width:

$$\left(\frac{\lambda}{d\lambda}\right)_{max} = \frac{2W_g}{\lambda}$$

For a given diffraction order, the grating resolution is constant across the working spectral range. A coarse grating with a few grooves designed to work in a high diffraction order may have the same resolution as a fine grating with a large number of grooves working in a low diffraction order.

Field Guide to Diffractive Optics

Grating Anomalies

Diffraction orders m satisfying the condition $|\sin\theta_m| < 1$ are called **propagating orders**. When varying the angle of incidence, some of the diffraction orders, called **passing-off orders** or **cut-off orders**, may propagate along the grating substrate. This is called the **threshold condition**:

$$\left|\frac{1}{n_2}\left(n_1 \sin\theta_i + \frac{m\lambda}{d_g}\right)\right| = 1$$

Small changes in the angle of incidence lead to the anomalies associated with the appearance or disappearance of the cut-off order and cause abrupt changes in the diffraction efficiency of the propagating orders. The disappearing orders that satisfy the condition $|\sin\theta_m| > 1$ are called **evanescent orders**.

Grating anomalies, also known as **Wood anomalies**, manifest as rapid variations in efficiency that occur within either narrow spectral or angular intervals. The wavelengths for the cut-off order are

$$\lambda = \frac{d}{m}[\text{sgn}(m) - \sin(\theta_i)]$$

The figure below shows the strong anomalous behavior of a sinusoidal reflective grating at a 30-deg angle of incidence for TM (S-polarized) light diffraction efficiency. The anomaly is caused by the emergence of the $m = -3$ order and the disappearance of the order $m = 1$.

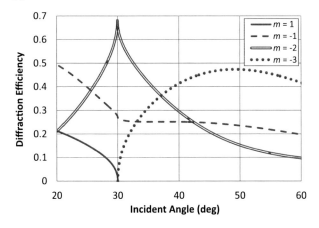

Polarization Dependency of Grating Anomalies

Rayleigh anomalies represent resonance phenomena due to the simultaneous occurrence of positive and negative diffractive orders propagating in opposite directions along the grating surface.

Grating anomalies have strong **polarization dependency** or **polarization anisotropy**. For reflective gratings, the anomalies are most prominent in transverse-magnetic or **TM-polarized** light. TM-polarized light has the electric field vector oriented parallel to the **incident plane**, also known as the **tangential plane**, and is therefore called *P*-polarized light. Anomalies for transverse-electric, or **TE-polarized** light, occurring when the electric vector is oriented in the **sagittal plane** (**S-polarized**) perpendicular to the incident plane, are observed in gratings with small groove spacing and deep grooves.

The first figure shows strong TM-polarized anomalies in the first diffraction order for a gold-coated grating at an incidence angle of 20 deg.

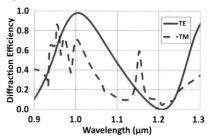

Grating anomalies can be observed in any diffraction order. The second figure shows 0th-order diffraction anomalies for a gold-coated grating at a 20-deg angle of incidence with two efficiency "notches" around 1.15 µm and 1.24 µm.

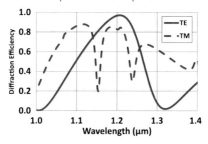

Field Guide to Diffractive Optics

Gratings as Angular Switches

Binary phase gratings can be employed to spatially redirect incoming radiation. The figure shows two binary gratings with a duty cycle of 0.5 laterally shifted by half of the grating period with respect to each other. The grating combination is equivalent to a plane-parallel plate, and the incoming radiation propagates unaltered through the grating pair.

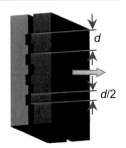

When the two gratings are relatively offset in the lateral direction by a quarter of the grating period, the grating combination becomes equivalent to a grating with triangular grooves. The incoming radiation is split into two beams at the exit of the grating pair.

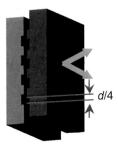

Resonance phenomena can be effectively employed to perform **angular switching** of radiation. The following figure presents angular switching between the zero and first diffraction orders of a reflective grating as a function of the incidence angle. By rotating the grating, the output of TM-polarized light can be rapidly switched between the two diffraction states.

A change in the grating angular orientation by 1 deg causes a change in the first diffraction order from an evanescent state to 75% output efficiency.

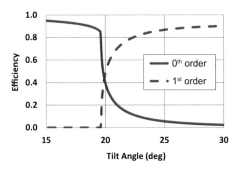

Field Guide to Diffractive Optics

Gratings as Optical Filters

The resonance phenomenon of grating anomalies can be effectively employed to alter grating properties. One practical outcome of using grating anomalies that is not obvious from grating behavior is the design of efficient **band-pass filters**.

The following figure shows the efficiency graph for TM-polarized light reflected into the first diffraction order with a full width at half maximum (FWHM) bandwidth of 0.54 µm centered at 1.6 µm. Reflection of TE-polarized light into the first diffraction order is also shown for comparison.

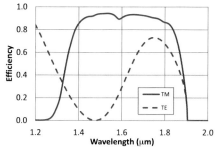

Resonance anomalies in subwavelength transmission gratings allow for the design of sharp notch filters working in the 0th diffraction order. The following graph shows a grating structure producing a narrow-band **etalon effect** in reflection to block the transmission of a laser beam at 1.06 µm, while transmitting neighboring wavelengths. The theoretical pass-band FWHM of the notch resonance is 0.14 nm.

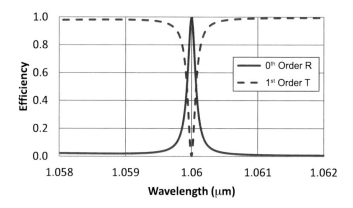

Field Guide to Diffractive Optics

Gratings as Polarizing Components

The design of reflective **diffractive polarizers** employs anomalous grating behavior. The figures below present a polarization grating performance designed at ~1.4 μm.

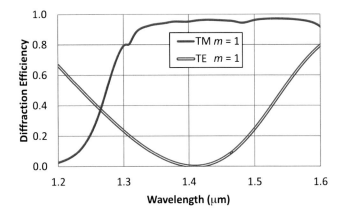

The grating effectively reflects TM polarization into the diffraction order $m = 1$ and reflects TE polarization into the diffraction order $m = 0$.

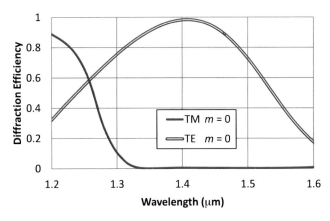

The **polarization extinction ratio** for TE-polarized radiation exceeds 200, and for TM-polarized radiation is greater than 1000.

Blazing Condition

Blazed gratings are composed of individual grooves that are shaped to concentrate the incident radiation into the diffraction order of interest.

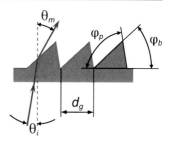

The **blazing condition** occurs when the propagation direction θ_m of the diffracted radiation coincides with the propagation direction through the grating microstructure defined by the laws of reflection or refraction. The blazing condition for a grating comprises grooves with a **blazed facet angle** φ_b is defined by a system of two equations:

$$\begin{cases} n_1 \sin(\theta_i + \varphi) = n_2 \sin(\theta_m + \varphi_b) \\ n_2 \sin(\theta_m) = n_1 \sin(\theta_i) + m\lambda/d_g \end{cases}$$

The angle of diffraction into the mth diffraction order θ_m can be found by

$$n_2 \cos(\theta_m) = n_1 \cos(\theta_i) - m\lambda/[d_g \tan(\varphi_b)]$$

The **blazing wavelength** λ_b is calculated as

$$\lambda_b = \frac{d_g}{m}\left(n_2 \sin\left\{\sin^{-1}\left[\frac{n_2}{n_1}\sin(\theta_i + \varphi_b)\right] - \varphi_b\right\} - n_1 \sin(\theta_i)\right)$$

The angle φ_p is the **passive facet angle**. For reflective gratings located in air, the blazing wavelength λ_b is found by

$$\lambda_b = \frac{2d_g}{m}\sin(\varphi_b)\sin(\theta_i + \varphi_b)$$

In **Littrow mounting**, the incident wave propagates at normal incidence to the grating facet, so that $\varphi_b = -\theta_i$, and the blazing condition becomes

$$\lambda_b^L = \frac{2d_g}{m}\sin(\varphi_b)$$

Field Guide to Diffractive Optics

Blazed Angle Calculation

The **blazed facet angle** at a given blazing wavelength λ_b, into a diffraction order m, and with a groove spacing d_g, is a function of the angle of incidence θ_i:

$$\tan(\varphi_b) = \frac{(m\lambda_b/d_g)}{n_1 \cos(\theta_i) - \sqrt{(n_2)^2 - [n_1 \sin(\theta_i) + (m\lambda_b/d_g)]^2}}$$

The following graph shows the blazed angle φ_b as a function of the incidence angle θ_i for different diffraction orders propagating from a medium with a lower refractive index into a medium with a higher refractive index.

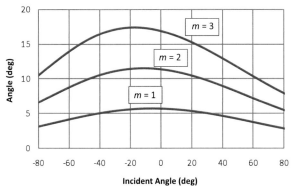

When propagating from a medium with a higher refractive index into a medium with a lower refractive index, the absolute blazed angle values φ_b as a function of the incident angle θ_i exhibit higher values:

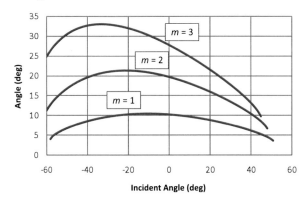

Field Guide to Diffractive Optics

Optimum Blazed Profile Height

The **optical path difference** ϕ between the neighboring facets of a blazed transmission grating produced as an interface between two materials with indices of refraction n_1 and n_2 is defined as

$$\phi = d_g[n_2\sin(\theta_m) - n_1\sin(\theta_i)] = m\lambda_b$$

The **optimum profile height** h_{opt} of a blazed grating surface operating in the mth diffraction order is derived as

$$h_{opt} = \left| m\lambda_b \middle/ \left\{\sqrt{(n_2)^2 - [n_1\sin(\theta_i) + (m\lambda_b/d_g)]^2} - n_1\cos(\theta_i)\right\}\right|$$

The first graph shows the optimum profile height as a function of the incidence angle for $m = 1$, $\lambda = 0.5$ µm, and three different grating periods when propagating from a less dense optical medium into a medium with a higher density when $n_1 < n_2$.

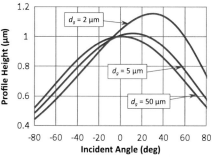

The optimum profile height as a function of the incidence angle for $m = 1$, $d_g = 10$ µm, and three different blazing wavelengths when $n_1 < n_2$ is shown in this graph:

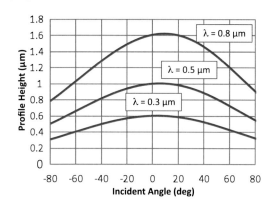

Field Guide to Diffractive Optics

Optimum Blazed Profile Height (cont.)

The first graph shows the **optimum profile height** h_{opt} for $m = 1$, $\lambda = 0.5$ µm, and three different grating periods when propagating from a higher to a lower density optical medium, such that $n_1 > n_2$.

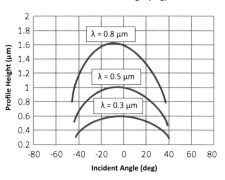

The optimum profile height as a function of the incidence angle for $m = 1$, $d_g = 10$ µm, and three blazing wavelengths is shown in the second graph.

The optimum profile height at a normal angle of incidence for a blazed grating in the mth diffraction order is simplified to

$$h_{opt} = \left| m\lambda_b \left/ \left[\sqrt{(n_2)^2 - (m\lambda_b/d_g)^2} - n_1 \right] \right. \right|$$

The optimum angle of the **passive facet angle** φ_p must be parallel to the direction of the diffracted field θ_m:

$$\varphi_p = \cos^{-1}\left[n_1 \sin(\theta_i)/n_2 + m\lambda_b/(n_2 d_g) \right]$$

This requirement cannot always be satisfied due to limitations associated with the grating profile fabrication processes.

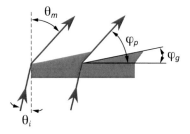

Field Guide to Diffractive Optics

Scalar Diffraction Theory of a Grating

Scalar diffraction theory enables an accurate definition of the propagation directions of diffracted light based on the grating equation. It also permits diffraction efficiency calculations when the grating period d significantly exceeds the operating wavelength λ. The gratings described by scalar diffraction theory have shallow facet angles. Therefore, the condition $d \gg h$ and the scalar diffraction theory are often referred to as the **thin-element approximation**. Diffractive optics that satisfy the condition $d \gg h$ belong to the **scalar domain**.

Scalar diffraction theory allows for the transition from a partial differential **wave equation** to an integral equation form. Diffraction efficiency in the mth transmitted order for gratings in the scalar domain can be defined as

$$\eta_m = \left| \frac{1}{d_g} \int_0^{d_g} t(x) e^{i\frac{2\pi}{\lambda}\left\{[n_1 \cos(\theta_i) - n_2 \cos(\theta_m)]f(x) - \frac{m\lambda}{d_g}x\right\}} dx \right|^2$$

where $f(x)$ is the grating profile function, $t(x)$ is the local transmission Fresnel coefficient, and x is the coordinate along the grating surface normal to the facets.

The relative grating facet size in the scalar domain is typically $d/\lambda \geq 15$. Polarization and shadowing effects on the grating facets are ignored in the thin-element approximation.

The grating period in general is a function of the lateral coordinates $d(x,y)$. The wavefront quality of a grating in the scalar domain is determined by the accuracy with which the spacing $d(x,y)$ is reproduced and is usually very good. In many cases the optical quality is limited by the quality of the substrate rather than by the errors associated with the groove spacing $d(x,y)$. Even for gratings with large apertures that have stitching errors, the wavefront quality is usually good, and the errors manifest in the presence of weak "ghost" orders.

Diffraction Efficiency

Diffraction efficiency in a given propagating order represents a fraction of the incident power contained within the order. The **relative diffraction efficiency** of a diffraction grating in the scalar domain is defined with respect to a perfectly reflecting or transmitting substrate. The diffraction efficiency η_m of an ideal blazed grating with blazing wavelength λ_b corresponding to the diffraction efficiency peak value depends on the operating wavelength λ and the diffraction order m:

$$\eta_m = \left\{ \sin\left[m\pi\left(\frac{\lambda_b}{\lambda} - 1\right) \right] \Big/ \left[m\pi\left(\frac{\lambda_b}{\lambda} - 1\right) \right] \right\}^2$$

The figure shows changes in diffraction efficiency as a function of the operating wavelength for diffraction gratings optimized at 500 nm and three different diffraction orders.

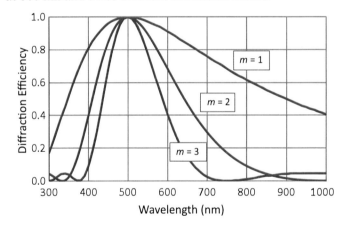

The normalized diffraction efficiency of a grating consisting of a finite number M of identical facets is a product of an interference term H from M beams and an intensity term I_d of a single facet with width d:

$$\eta_M = H I_S = \left[\frac{\sin\left(M \frac{kdp}{2}\right)}{M \sin\left(\frac{kdp}{2}\right)} \right]^2 \left[\frac{\sin\left(\frac{kdp}{2}\right)}{\frac{kdp}{2}} \right]^2$$

where $p = n_2 \sin\theta_m - n_1 \sin\theta_i$.

Field Guide to Diffractive Optics

Blaze Profile Approximation

The fabrication of high-fidelity triangular profile blazed gratings is costly and time consuming. Reduction in manufacturing costs is achieved by employing grating structures with approximated blazed profile shapes. **Lithographic techniques** are often employed to reduce grating fabrication costs.

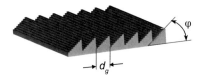

Binary gratings approximate a continuous grating profile with a staircase shape and are fabricated using a succession of lithographic etching steps. The diffraction efficiency η of a multilevel binary grating that employs N binary steps is calculated as

$$\eta = \left\{ \frac{\sin\left[m\pi\left(\frac{\lambda_b}{\lambda} - 1\right)\right]}{m\pi\left(\frac{\lambda_b}{\lambda} - 1\right)} \right\}^2 \left[\frac{\sin\left(\frac{\pi\lambda_b}{\lambda N}\right)}{\left(\frac{\pi\lambda_b}{\lambda N}\right)} \right]^2$$

The peak diffraction efficiency depends on the number of binary steps approximating the blazed profile:

Number of steps (N)	2	4	8	16	32	64
Peak Efficiency	40.5%	81.1%	95.0%	98.7%	99.7%	99.9%

Diffraction efficiency as a function of the operating wavelength for diffraction gratings optimized at 500 nm with differing numbers of binary steps is shown below:

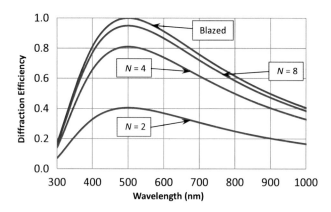

Field Guide to Diffractive Optics

Extended Scalar Diffraction Theory

Extended scalar diffraction theory improves the accuracy of efficiency calculations by accounting for the **shadowing effects**, the **fill factor** of the propagating field after diffraction (also called the **duty cycle** ζ of the diffracted field), and the **Fresnel reflections** at the substrate interfaces. The figure shows the appearance of gaps in the propagating wavefront and the reduction in fill factor after diffraction.

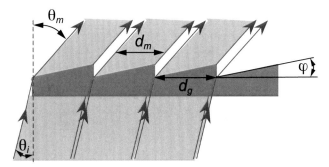

The field shadowing by the grating profile reduces the duty cycle ζ and diffraction efficiency. Changes in diffraction efficiency due to shadowing are accounted for by introducing the duty cycle ζ into the diffraction efficiency calculations:

$$\eta_{FF} \approx \zeta \eta_{scalar}$$

The duty cycle of a transmission grating is found by

$$\zeta = \frac{d_m}{d_g} = 1 - \frac{\left[n_1 \sin(\theta_i) + m\lambda/d_g\right]\tan(\varphi)}{\sqrt{(n_2)^2 - \left(n_1 \sin\theta_i + m\lambda/d_g\right)^2}}$$

Accounting for the blaze angle φ, the shadowing becomes

$$\zeta = 1 - \frac{1}{\sqrt{(n_2)^2 - \left(n_1 \sin\theta_i + m\lambda/d_g\right)^2}}$$

$$\times \frac{\left[n_1 \sin(\theta_i) + m\lambda/d_g\right](m\lambda/d_g)}{\left(\left|n_1 \cos(\theta_i) - \sqrt{(n_2)^2 - \left[n_1 \sin(\theta_i) + m\lambda/d_g\right]^2}\right|\right)}$$

Duty Cycle and Ghost Orders

When propagating from the substrate into air at normal incidence $\theta_i = 0$, the **duty cycle** due to shadowing is

$$\zeta = 1 - \frac{(m\lambda/d_g)^2}{\sqrt{(n_2)^2 - (m\lambda/d_g)^2}\left(\left|n_1 - \sqrt{(n_2)^2 - (m\lambda/d_g)^2}\right|\right)}$$

The influence of the shadowing on the diffraction efficiency of the propagating field can be explained by observing the far field of the diffracted radiation. Shadowing effects lead to the appearance of several secondary **ghost orders** in the vicinity of the primary diffraction order, as shown below for the duty cycle ζ ranging from 0.25 to 0.95.

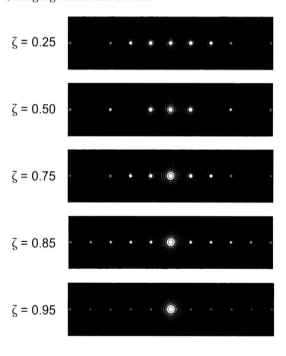

Extended Scalar versus Rigorous Analysis

Extended scalar diffraction theory provides an express calculation technique for the amount of light directed into different diffraction orders without the need for computationally extensive rigorous algorithms. The following graph compares the diffraction efficiencies of a blazed transmission grating in the first diffraction order as a function of the relative feature size d/λ for both the extended scalar and **rigorous diffraction analysis** techniques.

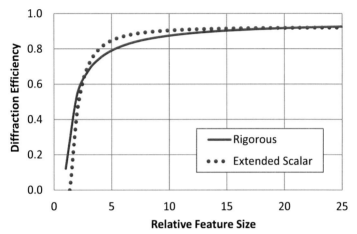

For a relative feature size of $d/\lambda > 15$, the difference in the diffraction efficiency calculated using both techniques is practically unnoticeable. For a feature size of $2.5 < d/\lambda < 15$, the diffraction efficiency calculated using the extended scalar theory provides optimistic efficiency results, while diffraction efficiency predictions for the feature size $d/\lambda < 2.5$ become overly pessimistic.

Diffraction efficiency of gratings with a feature size $d/\lambda < 15$ are approximated using the following equation:

$$\eta_{rig} \approx \eta_{scalar}[1 - C_m(n,\theta)\lambda_B/d]$$

where η_{scalar} is the maximum efficiency in scalar domain, and the coefficient $C_m(n,\theta)$ is a function of both the index of refraction and the angle of incidence.

Gratings with Subwavelength Structures

Advancements in high-resolution lithography and microfabrication have enabled diffractive structures with subwavelength feature sizes. As the period-to-wavelength ratio decreases, the diffraction efficiency predictions based on scalar diffraction become inaccurate. When the feature size is of the order of the wavelength or less, the scalar diffraction theory is no longer valid.

The design of subwavelength diffraction structures often employs rigorous electromagnetic propagation techniques that are based on the solution to **Maxwell's equations**. **Rigorous diffraction analysis** techniques can accurately account for the diffraction efficiency and polarization states of the propagating radiation.

In the case of both transmission and reflection gratings designed to blaze in TM polarization, the efficiency in the TE polarization state is always less than 100%. The blazing condition for both polarizations may occur when one of the polarization states propagates in transmission while the other state propagates in reflection.

These grating structures are employed as polarizing components in visible and IR photonics applications.

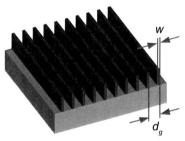

The size of the subwavelength grating period d_g is often selected to be smaller than the **structural cutoff** d_c, which is defined as the period below which all nonzero reflected and transmitted diffraction orders become evanescent. The necessary condition for the structural cutoff is

$$\frac{d}{\lambda} < \frac{1}{\sqrt{\max(n_1, n_2) + n_1 \sin(\theta_i)}}$$

Field Guide to Diffractive Optics

Blazed Binary Gratings

Due to practical limitations associated with mask misalignments and etching depth nonuniformity, an increased number of binary levels does not necessarily lead to improved grating performance. For example, a 32-level binary grating may have lower efficiency and higher scattering as compared to a 16-level grating.

Blazed binary grating profiles define surface-relief structures approximating the blaze profiles and requiring a single lithographic step for their fabrication. Blazed binary gratings have features smaller than the **structural cutoff**, defined as a feature size below which the grating behaves as a homogeneous layer.

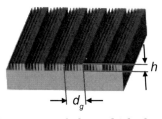

Blazed binary gratings represent **artificial dielectrics** and are designed using the **effective medium theory**. They operate in the **resonant domain**, where the grating period is equal to only a few wavelengths. Each grating period contains a series of ridges with continuously varying widths that change the local fill factor. When properly fabricated, the efficiency of the blazed binary gratings may exceed the efficiency of the blazed gratings operating in the scalar domain.

An alternate design for a blazed binary grating consisting of a number of individual "pillars" with progressively decreasing lateral dimensions is shown in the figure.

Field Guide to Diffractive Optics

Relative Feature Size in the Resonant Domain

The grating performance in the **resonant domain**, when the grating period d is comparable to the operating wavelength λ, rapidly changes with the **relative feature size** d/λ. In the case of a gold-coated reflection grating with a groove spacing of 1.42 µm and TE-polarized light propagating at a 20-deg angle of incidence, several diffraction orders coexist when the relative feature size $d/\lambda > 1.5$:

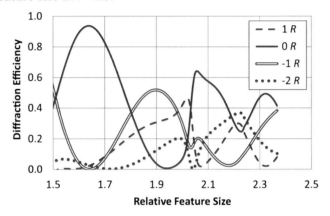

For the relative feature sizes $0.75 < d/\lambda < 1.5$, only the zero and negative first diffraction orders coexist, and the total reflected energy is redistributed between the two orders. For subwavelength feature sizes $d/\lambda < 0.75$, the negative first order vanishes, and the grating performs as a mirror.

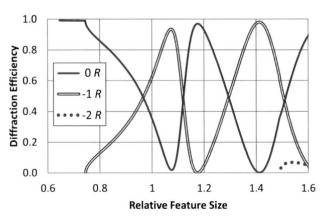

Effective Medium Theory

As shown earlier, for a gold-coated grating with subwavelength features, the structural cut-off condition for TE-polarized light is satisfied when $d_g/\lambda < 0.75$ and the grating surface behaves as a mirror. Similarly, for a transmission grating operating below the structural cutoff, all propagating diffraction orders except the zero order become evanescent. These types of gratings are called **zero-order gratings**. Zero-order gratings are employed as antireflection coatings, wave plates, and artificial distributed index media.

Based on **effective medium theory**, the effective index of refraction $n_\perp$ for a subwavelength **binary surface structure** and the electric field $\boldsymbol{E}_\perp$ orthogonal to the grating grooves is defined as

$$n_\perp = \frac{n_1 n_1}{\sqrt{(n_1)^2 \zeta_g + (n_2)^2 (1 - \zeta_g)}}$$

where n_1 and n_2 are the refractive indices of the grating substrate material and the medium surrounding the grooves, respectively; w is the groove width, and $\zeta_g = w/d_g$ is the grating duty cycle. The effective index of refraction $n_\parallel$ for the electric field $\boldsymbol{E}_\parallel$ parallel to the grating grooves is defined as

$$n_\parallel = \sqrt{(n_1)^2 \zeta_g + (n_2)^2 (1 - \zeta_g)}$$

The difference between the two effective indices of refraction $\triangle n = n_\perp - n_\parallel$ depends on the profile of the diffractive structure, which behaves as an **artificial dielectric** material that exhibits **form birefringence**.

The figure shows a grating profile with the effective index gradually changing over the grating period d_g.

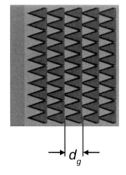

Scalar Diffraction Limitations and Rigorous Theory

Fresnel reflections at the air–grating interface cause the formation of forward- and backward-propagating diffraction orders, limiting the theoretically achievable diffraction efficiency in the blazed transmission order. In practice, uncoated transmission gratings transfer no more than 90% of the propagating radiation in the first diffraction order and about 80% in the second order. With a reduction in groove spacing, scalar theory becomes progressively less accurate, and rigorous simulation techniques are applied to optimize the grating structure.

Limitations of **scalar diffraction theory** become apparent when accurate efficiency calculations are required for gratings with smaller periods, or when changes in the polarization states of propagating light need to be accounted for. Larger discrepancies in efficiency predictions are observed for propagating light with the electric vector orthogonal to the grating grooves.

Rigorous diffraction analysis techniques produce accurate diffraction efficiency calculation results and are based on solving Maxwell's equations. Major rigorous techniques include **coupled wave analysis** based on space-harmonic expansion, **modal analysis** based on modal expansion, **finite difference methods**, as well as **integral methods**. With continuous advancements in computer speed, the **finite difference time domain** (FDTD) technique is establishing its place as a powerful practical technique for rigorous diffraction analysis.

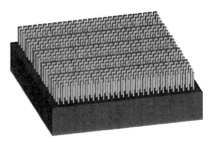

The figure shows a high-aspect-ratio grating structure requiring rigorous diffraction analysis techniques for performance evaluation.

Analysis of Blazed Transmission Gratings

Rigorous analysis of transmission gratings provides valuable insight into the physics of diffraction phenomena. The **diffraction efficiency** of **blazed transmission gratings** gradually decreases with a reduction in the grating period d_g. When the grating profile height h is set to be constant, the peak diffraction efficiency shifts toward the shorter wavelengths, as shown in the following figure for TM-polarized radiation in the first diffraction order $m = 1$:

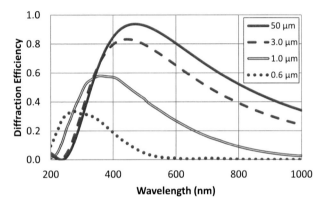

Even for grating feature sizes as small as 0.6 μm, the peak diffraction efficiency remains above 30%.

For the diffraction order $m = 2$, the respective efficiency graphs are shifted toward the shorter wavelength:

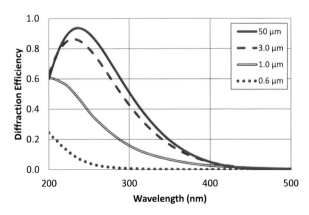

Polarization Dependency and Peak Efficiencies

For gratings with periods of $d_g \geq 1.0$ μm at normal incidence, the relative difference in the peak diffraction efficiencies for TM- and TE-polarization states does not exceed 2.4%. In general, at normal incidence a transmission grating diffraction efficiency has low **polarization dependency**, as shown in the following graph for the grating period $d_g = 3.0$ μm in the first diffraction order:

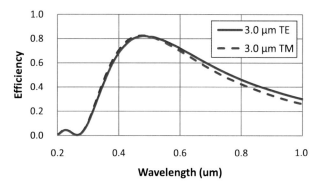

When the grating profile height is adjusted in accordance with the optimum height value h_{opt} defined based on extended scalar theory, the diffraction efficiency curves no longer shift toward the shorter wavelengths with a reduction in the grating period d_g. This is shown in the following figure for TM-polarized radiation:

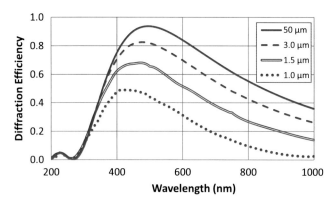

Field Guide to Diffractive Optics

Peak Efficiency of Blazed Profiles

Changes in the **peak diffraction efficiency** of a blazed transmission grating as a function of the feature size are shown in the next graph.

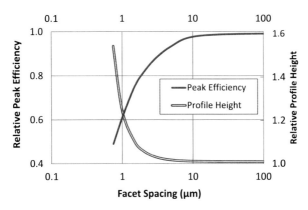

In the so-called **scalar domain**, when the relative feature size $d/\lambda > 10$, the peak diffraction efficiency variation is less than 2%.

Outside of the scalar domain, the diffraction efficiency of a blazed transmission grating rapidly degrades with a reduction in the relative feature size.

The peak diffraction efficiency and position of the peak wavelength as a function of the grating facet spacing change as follows:

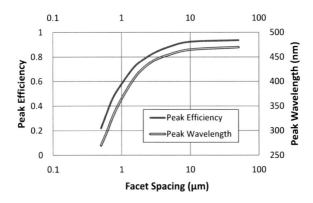

Field Guide to Diffractive Optics

Wavelength Dependency of Efficiency

The diffraction efficiency of blazed transmission gratings in a scalar domain ($d/\lambda > 10$) exhibits a characteristic red **spectral shift** as a function of the incidence angle. The figure shows the diffraction efficiency of a blazed transmission grating with feature size $d = 50$ µm and TM-polarized light at several angles of incidence as a function of the operating wavelength λ. The peak diffraction efficiency for TM-polarized radiation remains practically unchanged over a wide range of incidence angles (+/−60 deg).

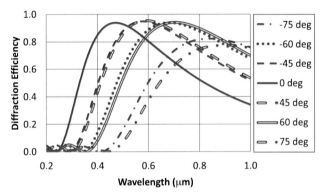

The peak diffraction efficiency for TE-polarized radiation gradually diminishes with an increase in the angle of incidence, as shown in this figure:

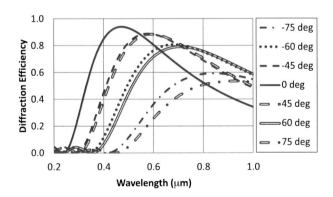

Efficiency Changes with Incident Angle

Diffraction efficiency as a function of the incident angle of TM-polarized light for blazed transmission gratings in a scalar domain (groove spacing $d = 50$) at several discrete wavelengths is shown in the following figure:

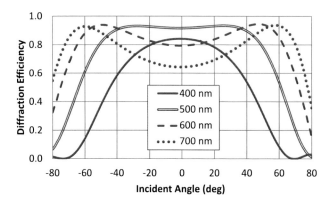

An increase in the operating wavelength above 400 nm leads to significant diffraction efficiency enhancements over the range of the incident angles, forming the batwing-shaped efficiency curves.

Diffraction efficiency curves may also exhibit **polarization anisotropy** at higher angles of incidence, as shown in the figure for a wavelength of 700 nm:

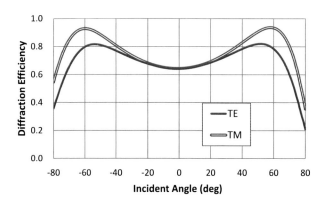

Field Guide to Diffractive Optics

Diffraction Efficiency for Small Feature Sizes

In the case of blazed gratings with small feature sizes ($d/\lambda < 15$), the facet angle is relatively steep, and the peak diffraction efficiency depends on both the angle of incidence and the wavelength of the incident light. The figure presents TM-polarized light diffraction efficiency for a blazed transmission grating with feature size $d = 3$ μm as a function of the operating wavelength λ:

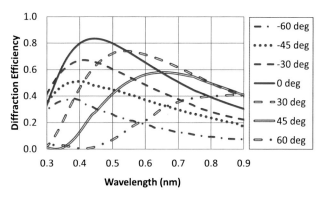

Diffraction efficiency curves exhibit asymmetry with respect to the angle of incidence, as shown for TM-polarized light diffracted by a blazed transmission grating with feature size $d = 3$ μm:

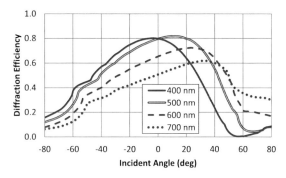

An increase in the operating wavelength leads to a peak **efficiency angular shift**.

Field Guide to Diffractive Optics

Polychromatic Diffraction Efficiency

The diffraction efficiency of a single grating surface (**diffractive singlet**) in a given diffraction order can be maximized only at a single operating wavelength, where the in-phase condition for the individual diffracted fragments of the propagating wavefront is satisfied.

An appropriately designed **diffractive doublet** based on the combination of two grating surfaces can produce high **polychromatic diffraction efficiency** over an extended operating spectral range, known as **broadband blazing**, shown for the wavelength range 350 nm < λ < 650 nm:

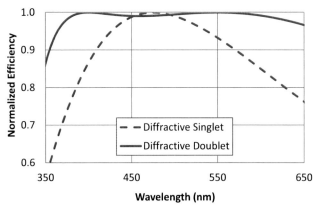

The polychromatic diffraction efficiency of the **grating doublet** shown in the graph is 98.7% over the spectral range from 350 to 650 nm, as compared to the polychromatic diffraction efficiency of 88.8% for a single diffraction structure with efficiency maximum at 480 nm.

The employment of diffractive singlets in imaging systems operating over an extended spectral range offers significant benefits in correcting the system's aberrations. These benefits are often outweighed by the image contrast degradation associated with reduced polychromatic diffraction efficiency of a diffractive singlet. Diffractive doublets can provide similar aberration correction benefits with higher image contrast.

Monolithic Grating Doublet

In the case of a **diffractive singlet**, the blazing condition is satisfied only for a single wavelength λ_b, and the polychromatic diffraction efficiency is reduced.

The **broadband blazing** condition can be achieved by employing a combination of two diffractive structures that form a **grating doublet**, either monolithic or spaced. A **monolithic grating doublet** is shown in the figure.

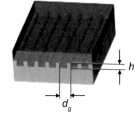

The **optical path difference (OPD)** introduced by a monolithic grating doublet with spacing d_g and step height h made of optical materials with refractive indices $n_1(\lambda)$ and $n_2(\lambda)$ is found based on the blazing condition

$$OPD = m\lambda_b = h\,[n_2(\lambda)\cos(\theta_m) - n_1(\lambda)\cos(\theta_i)]$$

The broadband blazing condition requires the OPD to be proportional to the individual wavelengths over the extended spectral range. Provided that the incident angle does not change, the broadband blazing condition for the monolithic grating doublet is defined as

$$\frac{d}{d\lambda}[n_2(\lambda)]\cos(\theta_m) - \frac{d}{d\lambda}[n_1(\lambda)]\cos(\theta_i) = m$$

Grating doublets can be made with various profiles, including triangular, lamellar, sinusoidal, etc.

Triangular profile

Lamellar profile

While monolithic grating doublets are relatively insensitive to fabrication errors, the broadband blazing condition for a monolithic grating doublet is difficult to satisfy due to the limited choice of materials satisfying the blazing requirements.

Field Guide to Diffractive Optics

Spaced Grating Doublet

Material constraints are greatly relaxed for **spaced grating doublets**. The OPD introduced by grating step heights h_1 and h_2 made of optical materials with refractive indices $n_1(\lambda)$ and $n_2(\lambda)$ and a spacer with index $n_3(\lambda)$ is defined as

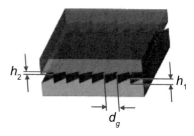

$$OPD = m\lambda_{bl} = h_1[n_3(\lambda)\cos(\theta_{m1}) - n_1(\lambda)\cos(\theta_{i1})] + \\ h_2[n_2(\lambda)\cos(\theta_{m2}) - n_3(\lambda)\cos(\theta_{i2})]$$

Broadband blazing for the spaced grating doublet is defined as

$$h_1\left\{\frac{d}{d\lambda}[n_3(\lambda)]\cos(\theta_{m1}) - \frac{d}{d\lambda}[n_1(\lambda)]\cos(\theta_{i1})\right\} + \\ h_2\left\{\frac{d}{d\lambda}[n_2(\lambda)]\cos(\theta_{m2}) - \frac{d}{d\lambda}[n_3(\lambda)]\cos(\theta_{i2})\right\} = m$$

For an **air-spaced grating doublet** with $d[n_3(\lambda)]/d\lambda \cong 0$, the broadband blazing condition is reduced to

$$\frac{d}{d\lambda}[n_2(\lambda)]h_2\cos(\theta_{m2}) - \frac{d}{d\lambda}[n_1(\lambda)]h_1\cos(\theta_{i1}) = m$$

The broadband blazing condition is satisfied for at least two operating wavelengths. A grating doublet significantly extends the broadband diffraction efficiency as compared to a grating singlet:

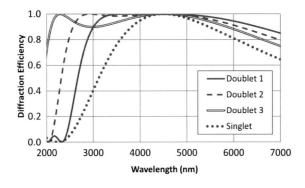

Monolithic Grating Doublet with Two Profiles

In order to eliminate the material constraints associated with **monolithic grating doublets** while maintaining the low doublet sensitivity to variations in the profile height, a second diffraction profile is formed at the doublet exterior. The second exterior profile can be formed, for example, by compression molding or etching.

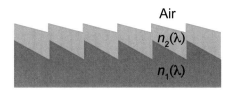

A dual-profile monolithic grating doublet has a significantly lower sensitivity to fabrication errors as compared to an air-spaced grating doublet. The depth errors Δh_1 in the grating substrate structure with refractive index $n_1(\lambda)$ lead to OPD errors proportional to the refractive index difference between the two doublet materials:

$$\Delta OPD = \Delta h_1 [n_1(\lambda) - n_2(\lambda)]$$

These OPD errors are several times smaller than similar errors of an air-spaced doublet, which are proportional to the difference between the refractive index of the substrate material and that of air:

$$\Delta OPD = \Delta h_1 [n_1(\lambda) - 1]$$

The exterior profile of the monolithic grating doublet is usually shallower than the substrate profile(s) employed in air-spaced grating doublets.

The dual-profile monolithic grating doublet provides an additional benefit of reduced Fresnel losses due to a reduced number of optical interfaces.

Diffractive and Refractive Doublets: Comparison

There is a similarity between diffractive and **refractive doublets**. Both types of doublets are intended to reduce the spectral dependency of one of the key performance parameters. In the case of a diffractive doublet, this parameter is the **polychromatic diffraction efficiency**, while in the case of a refractive doublet, it is the size of the **polychromatic point spread function**.

The net **refractive optical power** of a refractive doublet is the sum of the positive and the negative optical powers of the two refractive elements composing the refractive doublet. The net optical power of a refractive doublet is less than the optical power magnitude of the two lens components.

Diffractive optical power is proportional to the spectral dispersion of the propagating field. For the diffractive doublet, the net diffractive power is the difference of the powers of the two diffractive components. The net diffractive power in the diffractive doublet with high polychromatic diffraction efficiency is lower than the powers of the individual diffractive components.

The axial thickness of a refractive doublet and the sensitivity of its components to surface shape variations significantly exceed those of a refractive singlet of equal optical power.

The diffractive profile heights and the sensitivity to profile height variations of the doublet grating components considerably exceed those of a diffractive singlet of equal diffractive power.

Surface profile errors in the refractive components cause distortions that lead to an increase in the refractive doublet PSF. Surface profile height errors in diffractive doublets lead to reduced polychromatic diffraction efficiency in the working order.

Efficiency of Spaced Grating Doublets

The diffraction efficiency of a **spaced grating doublet** is a function of several parameters, including polarization, wavelength, angle of incidence, material properties, step heights, and the grating spacing.

In most cases, a rigorous diffraction analysis technique is required to design and optimize the performance of grating doublet structures. The following analysis shows the performance of an air-spaced grating doublet designed to operate in the visible spectrum and made of two dissimilar materials with refractive indices $n_1 = 1.4623$ and $n_2 = 1.6588$. Diffraction efficiency exhibits a strong asymmetric angular dependency, as shown in the following graphs for TE-polarized light and positive and negative angles of incidence.

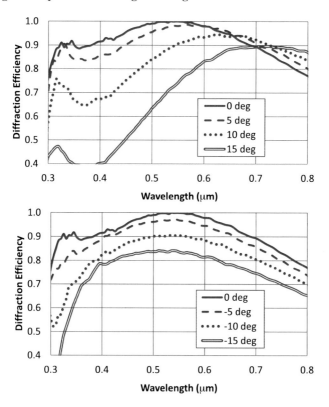

Field Guide to Diffractive Optics

Efficiency of Spaced Grating Doublets

Sensitivity to Fabrication Errors

The fabrication of spaced grating doublets with **broadband diffraction efficiency** requires a relatively high fabrication accuracy of individual grating components constituting the doublet. The following graphs show the grating doublet diffraction efficiency sensitivity to changes in step height of one of the grating components for TE- and TM-polarized light, respectively.

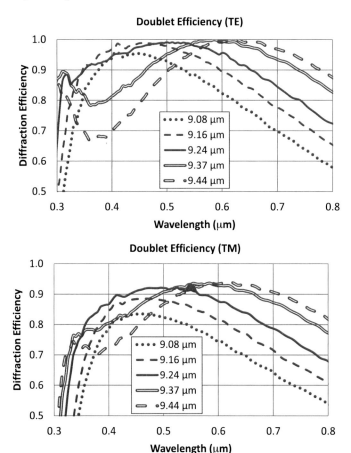

Facet Width and Polarization Dependency

The **broadband diffraction efficiency** of air-spaced grating doublets depends on the **facet width** of the gratings composing the doublet. For grating doublets with a broadband diffraction efficiency optimized over the visible range, significant efficiency degradation is observed for the facet widths of less than 50 µm.

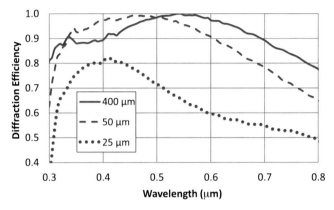

The broadband diffraction efficiency of air-spaced grating doublets has a significant **polarization dependency**. TE-polarized light has higher diffraction efficiency than TM-polarized light over the operating spectral range. This can be explained qualitatively by the differences in Fresnel reflection losses of the two polarization states.

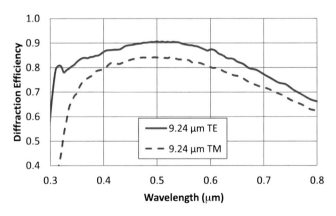

Field Guide to Diffractive Optics

Sensitivity to Axial Component Spacing

Polychromatic diffraction efficiency of an air-spaced diffraction doublet depends on the axial spacing between the doublet components. Changes in axial spacing t between the diffraction doublet components cause lateral shifts of the propagating field fragments diffracted by the first grating with respect to the facets of the second grating.

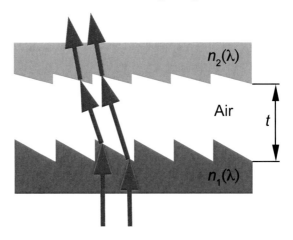

An increase in the axial spacing t between the two grating components leads to an oscillatory spectral response of the doublet. The **polychromatic efficiency modulation** increases with an increase in the axial spacing t, as shown in the figure for TE-polarized light.

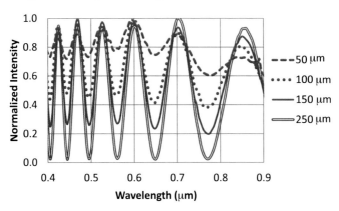

Field Guide to Diffractive Optics

Frequency Comb Formation

The lateral shift of the field diffracted by the first grating in the doublet leads to wavefront division by the facets of the second grating. Interference of the wavefront fragments after the second grating causes formation of a **frequency comb** in the spectral response of the doublet, as shown in the figure for two different values of the axial spacing.

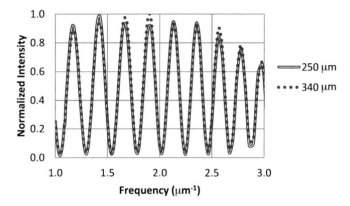

Nodal locations in the transmitted frequency comb do not change with adjustments in the axial separation between the doublet components. The oscillatory spectral response of the grating doublet is observed in different diffraction orders. The figure shows alternating extrema locations for the spectral efficiency of TE-polarized light diffracted by the doublet with an air gap of 250 μm into the orders 0, 1, and 2.

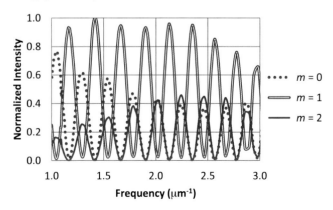

Field Guide to Diffractive Optics

Diffractive Components with Axial Symmetry

Diffractive components with axial symmetry constitute one of the largest groups of diffractive structures and include amplitude masks, phase plates, diffractive lens surfaces, stepped diffractive surfaces (SDSs), hybrid diffractive-refractive structures, as well as lens diffractive doublets. Axially symmetric diffractive components are found in a variety of applications, including imaging objectives operating in the UV, visible, and IR, as well as intra-ocular lenses for vision correction and in illumination and laser systems.

Amplitude masks with axial symmetry, such as Fresnel zone plates, are based on amplitude modulation of the incident radiation. Amplitude masks are employed in the spectral regions where nonabsorbing optical materials are not readily available, such as in the extreme UV. Employment of amplitude masks is associated with net transmission losses introduced by the masks.

Annular phase plates, such as Fresnel phase plates, are based on the phase modulation of the incident wavefront by introducing phase delay to the different portions of the propagating wavefront. Phase plates are designed to work in **transmission** or **reflection**, or to produce **bidirectional propagation**. With an appropriate choice of coatings and phase plate material, minimal transmission losses are introduced into the system.

Diffractive lens surfaces include kinoform, binary, and multi-order diffractive lenses and represent phase structures that contribute to the net optical power of their optical systems.

Stepped diffractive surfaces (SDSs) represent diffractive phase structures with axially symmetric zones and zero optical power at the design wavelength. An SDS cross section is shaped like a staircase. Each SDS zone produces a phase delay that is an integer multiple of 2π.

Diffractive Lens Surfaces

Diffractive lens surfaces (DLSs) are composed of annular grooves and contribute to the net **optical power** in an optical system. The group of DLSs includes **diffractive kinoforms** and **binary surface structures**, as well as refractive-diffractive and reflective-diffractive **hybrid structures** fabricated into a curved substrate.

Diffractive kinoforms are grating phase structures consisting of circular phase zones with radially varying **blazed facet angles**. Diffractive kinoforms are based on the phase modulation of the incident wavefront by introducing a radially variable **phase delay** to the different portions of the propagating wavefront.

Longer wavelengths focus closer to the diffractive surface.

Locations of the focal points for three wavelengths

$\lambda_r > \lambda_g > \lambda_b$

satisfy the inequality

$f_r < f_g < f_b$

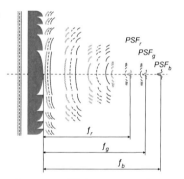

Binary surface structures are made as multistep approximations to the kinoform zone profiles. Each **binary level** produces a phase delay that is a fraction of a 2π kinoform zone delay (usually ranging from $\pi/32$ to π).

The hybrid refractive-diffractive and reflective-diffractive surface structures combine the properties of diffractive kinoforms with the refractive or reflective properties of the substrate.

Diffractive lens surfaces are designed for use in imaging and nonimaging applications, including illumination and laser optics.

Field Guide to Diffractive Optics

Diffractive Kinoforms

Diffractive kinoforms are surface-relief structures fabricated on a planar substrate that comprise several concentric zones with continuous profiles. The term "kinoform" was introduced in 1969 by researchers from IBM. The **optical power** of a kinoform is proportional to the number of zones N_k and the design wavelength λ_0 and is inversely proportional to the square of the clear aperture diameter D_0:

$$\Phi^D(\lambda) = 8N_k\lambda_0/(D_0)^2$$

A diffractive kinoform resembles the shape of a conventional **Fresnel lens**, and its cross section looks like a series of sawtooth-shaped ridges with variable radial spacing d_i and facet angle φ_i. In contrast with Fresnel lenses, diffractive kinoforms precisely control the OPD introduced by the zones and are composed of a significantly larger number of zones.

The OPD ϕ between two neighboring kinoform zones is a product of the diffraction order m and the **design wavelength** λ_0:

$$\phi = m\lambda_0$$

For a given design wavelength λ_0 and refractive index n_0 of the substrate, the **optimum zone height** h of the diffractive kinoform operating in the first diffraction order $m = 1$ is a function of the incident angle θ_i:

$$h(\theta_i) = \frac{\lambda_0}{\sqrt{(n_0)^2 - (\sin\theta_i)^2} - \cos\theta_i}$$

The kinoform zone thickness profile $t(r, \theta_i)$ is found by

$$t(r, \theta_i) = h(\theta_i)(\phi(r) \bmod 2\pi)$$

For on-axis propagation ($\theta_i = 0$), the zone height is

$$h_0 = \lambda_0/(n_0 - 1)$$

Binary Diffractive Lenses

Binary diffractive lenses are surface-relief structures that are fabricated based on **very large-scale integration (VLSI)** lithographic techniques. The name "binary" reflects the coding scheme employed to create the sequence of photolithographic masks for fabrication. Binary lens coding provides a step-wise approximation to the radial phase profile $\psi(r)$ of a diffractive kinoform.

The figure compares a three-zone diffractive kinoform and the respective three-zone binary lens with four binary levels approximating the diffractive kinoform.

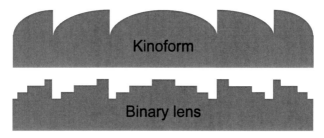

To reduce the number N of processing cycles and the respective number of masks required during binary lens fabrication, a binary coding scheme is used. The number of binary levels is defined as 2^N, and the thickness t_b of each binary lens level for the design wavelength λ_0, zone height h_0, and refractive index n_0 is found by

$$t_b = \frac{h_0}{2^N} = \frac{\lambda_0}{(n_0 - 1)2^N}$$

The diffraction efficiency of binary lenses is less than the efficiency of the respective kinoforms. Efficiency reduction is caused by profile approximations, additional losses associated with interfacial roughness, and fabrication imperfections of the binary steps. Increasing the number of binary steps to greater than 32 may not necessarily produce a binary lens with higher efficiency, due to increased interfacial roughness, etch depth variations, and mask misalignments.

Optical Power of a Diffractive Lens Surface

The **optical power** Φ^D of a diffractive lens surface at any operating wavelength λ is related to the nominal optical power Φ_0^D at the design wavelength λ_0 as

$$\Phi^D(\lambda) = \Phi_0^D \frac{\lambda}{\lambda_0}$$

The focal plane locations for the wavelengths

$$\lambda_r > \lambda_g > \lambda_b$$

satisfy the inequality

$$f_r < f_g < f_b$$

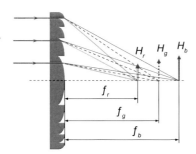

The difference $\Delta f_{chr} = f_b - f_r$ is called **longitudinal** or **axial chromatic aberration**. The change in image size at the different wavelengths $\Delta H_{chr} = H_b - H_r$ is called **transverse** or **lateral chromatic aberration**.

Axial chromatic aberration of a diffractive surface Δf_{chr}^D is proportional to the **spectral bandwidth** $\Delta\lambda = \lambda - \lambda_0$ of the propagating field and can be expressed as

$$\Delta f_{chr}^D = \Delta\lambda / \left(\Phi_0^D \lambda_0\right)$$

For comparison, the optical power Φ^R of a refractive surface can be defined as

$$\Phi^R(\lambda) = \frac{n(\lambda) - 1}{R_0} = \Phi_0^R \left[1 + \frac{D_n(\lambda)\Delta\lambda}{n_0(\lambda) - 1}\right]$$

where $\Phi_0^R = [n_0(\lambda) - 1]/R_0$ is the nominal refractive surface power corresponding to the substrate radius R_0 with refractive index n_0 at the design wavelength λ_0. The lens **material dispersion** $D_n(\lambda)$ is calculated as

$$D_n(\lambda) = \frac{n(\lambda) - n_0(\lambda)}{\lambda - \lambda_0}$$

Diffractive Surfaces as Phase Elements

For raytracing purposes, it is convenient to represent diffractive surfaces as phase elements with respective **phase profiles** $\psi(x,y)$ that depend on the surface lateral coordinates x and y.

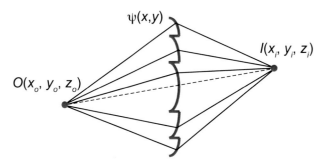

A phase profile $\psi(x,y)$ of a diffractive lens surface forming an image $I(x_i, y_i, z_i)$ of a point $O(x_o, y_o, z_o)$ in the object space at the design wavelength λ_0 is defined as

$$\psi(x,y) = \frac{2\pi}{\lambda_0} \left\{ \left[\sqrt{(x-x_o)^2 + (y-y_o)^2 + (z_o)^2} - z_o \right] - \left[\sqrt{(x-x_i)^2 + (y-y_i)^2 + (z_i)^2} - z_i \right] \right\}$$

The axial distances z_o and z_i satisfy the sign convention and are positive to the right of the surface. The optical power $\Phi^D(\lambda)$ of the surface is then given by the **Gaussian lens formula**

$$\Phi^D = 1/z_i - 1/z_o$$

In many practical applications, including aberration correction, beam shaping, and athermalization, the diffractive surface does not form images of an object. For surfaces without rotational symmetry, the phase is

$$\psi(x,y) = m \sum_{k=0}^{K} \sum_{l=0}^{L} A_{kl} (x)^k (y)^l$$

For axially symmetric diffractive surfaces, the phase is a function of the radial coordinate r and order m:

$$\psi(r) = m \sum_{i=0}^{N} A_i (r)^{2i}$$

Stepped Diffractive Surfaces

Stepped diffractive surfaces (SDSs) are powerless surface-relief diffractive phase structures fabricated as a set of concentric circular zones with planar interfaces, also referred to as "steps." Components employing SDSs are sometimes called **staircase lenses**.

The steps are bounded by cylindrical sections. Each zone is characterized by a width d_i and is axially offset from the neighboring zone by a distance h_i.

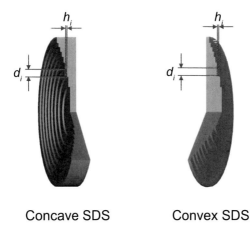

Concave SDS   Convex SDS

The nominal **optical power** of an SDS at the design wavelength λ_0 is zero. Therefore, SDSs can be added to optical systems without altering the first-order properties.

The **shadowing effects** produced by stepped diffractive surfaces at the zone boundaries are significantly smaller as compared to diffractive lens surfaces with the same apertures and dispersion values, leading to increased diffraction efficiency.

While complex diffractive lens profiles require radially varying facet angles with sharp corners that are difficult to reproduce, SDS staircase profiles are easier to fabricate using direct **single-point diamond turning** (SPDT) or replication techniques.

Field Guide to Diffractive Optics

Properties of Stepped Diffractive Surfaces

The **optical power** of an SDS depends on the operating wavelength λ and the refractive indices $n_1(\lambda)$ and $n_2(\lambda)$ of the materials before and after the interface, respectively:

$$\Phi^{SDS}(\lambda) = \left| \frac{n_2(\lambda) - n_1(\lambda)}{R_0} \right| \left(\frac{\lambda}{\lambda_0} - 1 \right)$$

SDS optical power at the **design wavelength** λ_0 is zero.

SDSs have several advantages over blazed diffractive kinoforms and their binary approximations:

1. An SDS can be added to any optical system without changing the paraxial optical power of the system. It can be employed in an optical system to correct for chromatic and/or monochromatic aberrations or to provide passive **athermalization** of the system.

2. The simple staircase zone geometry of an SDS allows for accurate microstructure fabrication by using the SPDT technique, leading to reduced aberrations caused by fabrication errors as compared to a diffractive lens surface with an equal aperture and power.

3. Compared to a binary diffractive surface, an SDS does not involve an approximation of the microstructure's shape and therefore produces higher diffraction efficiencies than binary diffractive lenses.

The optimum step height h_{SDS} of an SDS is a function of the incident angle θ_i, the working diffraction order m, and the design wavelength λ_0, and is found by

$$h_{SDS} = \left| \frac{m\lambda_0}{\sqrt{(n_2)^2 - [\sin(\theta_i)]^2} - n_1 \cos(\theta_i)} \right|$$

In the case of a normal-incidence SDS, the step height becomes

$$h_{SDS} = \left| \frac{m\lambda_0}{n_2 - n_1} \right|$$

Multi-order Diffractive Lenses

Multi-order diffractive lenses are designed to produce OPDs between the zones that are a multiple m of the operating wavelengths. Multi-order diffractive lenses are similar to **echelle** diffractive gratings with high-profile depths and are designed to work in high diffraction orders. For a multi-order diffractive lens designed to operate at an infinite conjugate in the mth diffraction order, the profile height can be defined as

$$h_m = \frac{m\lambda_0}{n(\lambda_0) - 1}$$

Multi-order diffractive lenses are often designed to blaze in more than a single operating wavelength. For a multi-order lens designed to operate at wavelengths λ_B and λ_R in respective orders m and k, the blaze condition becomes

$$\frac{m\lambda_B}{n(\lambda_B) - 1} = \frac{k\lambda_R}{n(\lambda_R) - 1}$$

The above blaze condition may be difficult to satisfy while achieving reasonably shallow depths of the diffractive zones. In that case, the profile depth is designed to blaze at the shorter wavelength λ_B. The graph presents changes in the diffraction efficiency for a multi-order diffraction lens designed to operate in the order $m = 3$ for the visible wavelength $\lambda_B = 0.532$ µm, as well as in the order $k = 1$ at a longer infrared wavelength $\lambda_R = 1.55$ µm.

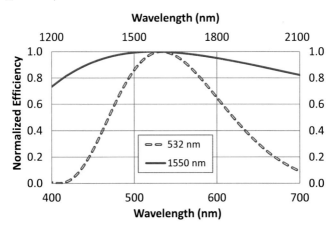

Field Guide to Diffractive Optics

Diffractive Lens Doublets

High diffraction efficiency for a single diffractive surface is achieved only over a narrow spectral bandwidth $\Delta\lambda$. In applications with broad radiation bandwidth, a significant fraction of radiation is directed into the **spurious diffraction orders** and causes a degradation of the system performance by reducing the system's resolution and image contrast, as well as by increasing the crosstalk in the detection channels.

Increased polychromatic diffraction efficiency is achieved by designing **diffractive lens doublets** consisting of two diffractive surfaces with opposite powers. The lens doublets are made as monolithic or air spaced:

Monolithic diffractive doublet   Air-spaced diffractive doublet

The following figure shows theoretical diffraction efficiencies for three air-spaced diffractive lens doublets made of different material pairs and designed to operate over an extended spectral range in the visible spectrum. All three doublets have efficiency maxima at 400 and 550 nm. The material choice plays an insignificant role in the achieved diffraction efficiency. Instead, other factors such as manufacturability, material cost, and required etch depth of the phase zones play dominant roles.

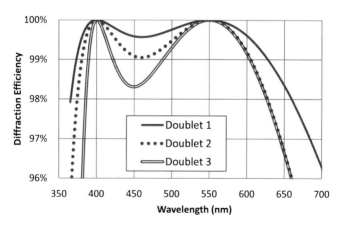

Field Guide to Diffractive Optics

Diffractive Lens Surfaces in Optical Systems

The small surface-relief thickness of diffractive lens surfaces makes them well suited for integration into optical systems. A surface-relief diffractive structure can be combined with a refractive or reflective surface to attain additional degrees of freedom in optical design.

Hybrid components combine diffractive surfaces with refractive or reflective counterparts. **Hybrid structures** integrate diffractive structures into refractive or reflective surfaces.

The total optical power of the hybrid structure Φ^H containing diffractive as well as reflective or refractive surfaces is found as the sum of the diffractive power Φ^D and the respective substrate power Φ^R:

$$\Phi^H(\lambda) = \Phi^R(\lambda) + \Phi^D(\lambda) = \Phi_0^R \left[1 + \frac{D_n(\lambda)\Delta\lambda}{n_0(\lambda) - 1} \right] + \Phi_0^D \left(1 + \frac{\Delta\lambda}{\lambda_0} \right)$$

Material dispersion $D_n(\lambda)$ depends on the operating spectral range $\lambda - \lambda_0$ as well as on the material's refractive index change with wavelength, as shown below:

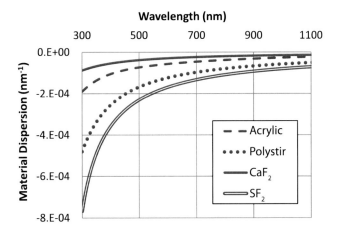

Achromatic Hybrid Structures

An **achromatic condition** for a hybrid doublet structure $\Delta\Phi^H = \Phi^H(\lambda) - \Phi^H(\lambda_0) = 0$ is satisfied when the relative power of the refractive surface equals

$$\frac{\Phi_0^R}{\Phi_0^D} = \frac{\Delta\lambda[n_0(\lambda) - 1]}{\lambda_0[n_0(\lambda) - n(\lambda)]} = -\frac{[n_0(\lambda) - 1]}{\lambda_0 D_n(\lambda)}$$

The relative power of the refractive surface required to satisfy the achromatic condition $\Delta\Phi^H = 0$ is shown in the graph for different substrate materials:

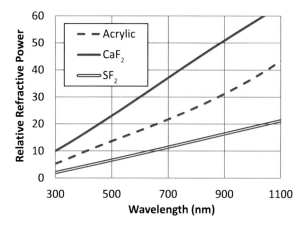

The relative diffractive power in a **hybrid achromat** is reduced with an increase in the operating wavelength.

The total power of the **achromatic hybrid structure** Φ^{AH}, found as the sum of the refractive and diffractive powers, is higher than the power of the diffractive Φ_0^D or refractive Φ_0^R surfaces of the doublet:

$$\Phi^{AH} = \Phi_0^D \left\{ 1 - \frac{[n_0(\lambda) - 1]}{\lambda_0 D_n(\lambda)} \right\} = \Phi_0^R \left\{ 1 - \frac{\lambda_0 D_n(\lambda)}{[n_0(\lambda) - 1]} \right\}$$

In the case of an **achromatic refractive doublet** the two refractive lenses have opposite powers, and the net optical power of a **refractive achromat** is reduced. Therefore, hybrid achromats yield lower monochromatic aberrations, axial thickness, and weight than respective refractive achromats.

Field Guide to Diffractive Optics

Diffractive Surfaces in Optical Systems

Opto-thermal Properties of Optical Components

The thermal behavior of an optical surface can be described in terms of an **opto-thermal coefficient** (**OTC**) ξ. An OTC defines the thermally induced relative rate of change in optical power over the temperature range ΔT:

$$\xi = \Delta\Phi/(\Phi\Delta T)$$

Thermally induced changes in the kinoform optical power are caused by changes in the zone spacing. Respective changes in the substrate refractive index affect only the diffraction efficiency of the lens.

The OTC of a kinoform ξ_K depends only on the substrate **coefficient of thermal expansion** (**CTE**) α:

$$\xi_K = -2\alpha$$

Changes in the optical power of a refractive interface are due to changes in the surface radius and in the substrate refractive index. The OTC of a refractive interface ξ_R is calculated as

$$\xi_R(\lambda) = \frac{1}{[n_2(\lambda) - n_1(\lambda)]} \left[\frac{dn_2(\lambda)}{dT} - \frac{dn_1(\lambda)}{dT} \right] - \alpha$$

Changes in the optical power of an SDS are due to changes in the zone spacing and in the refractive indices $n_1(\lambda)$ and $n_2(\lambda)$ of the media before and after the interface, respectively. The opto-thermal coefficient of an SDS depends on both the CTE of the substrate material and the thermally induced changes in the refractive indices dn/dT:

$$\xi_{SDS}(\lambda) = \frac{1}{[n_2(\lambda) - n_1(\lambda)]} \left[\frac{dn_2(\lambda)}{dT} - \frac{dn_1(\lambda)}{dT} \right] + \alpha$$

The OTC sign of an SDS is positive when the SDS is convex and negative when the SDS is concave.

The opto-thermal coefficients of an SDS, a refractive surface, and a diffractive kinoform are related as follows:

$$\xi_{SDS}(\lambda) = \pm[\xi_R(\lambda) - \xi_K]$$

Field Guide to Diffractive Optics

Athermalization with Diffractive Components

Thermally induced relative changes in power Φ of an optical surface are proportional to the temperature changes ΔT and the surface opto-thermal coefficient ξ:

$$\frac{\Delta \Phi}{\Phi} = \xi \Delta T$$

Diffractive components provide additional degrees of freedom in designing **athermal lens** solutions. During athermal lens design it is also necessary to account for the optical component mounting scheme and the relative component shifts based on the CTE of the housing material.

For an optical system containing N optical interfaces, the **athermal condition** occurs when the net change in optical power of the lens system, defined based on the contribution $\Delta \Phi_i$ from all of the optical interfaces, is zero:

$$\sum_{i=1}^{N} (\Delta \Phi_i) = 0$$

The athermal condition for a hybrid refractive–diffractive kinoform singlet lens and for thermally invariant housing can be written as

$$\xi_K \Phi^K = -\xi_R \Phi^R$$

The ratio of the optical powers of the kinoform Φ^K and the refractive Φ^R surfaces in the athermal singlet is

$$\frac{\Phi^K}{\Phi^R} = \frac{\left\{ \frac{d}{dT}[n_2(\lambda)] - \frac{d}{dT}[n_1(\lambda)] \right\}}{2\alpha[n_2(\lambda) - n_1(\lambda)]} - \frac{1}{2}$$

An **athermal achromat** can be constructed based on a single hybrid lens element and an appropriate selection of mounting material.

Field Guide to Diffractive Optics

Athermalization with SDSs

Opto-thermal coefficients for refractive, diffractive, and SDS surfaces depend on the substrate material.

Material	$\xi_R \times 10^{-6}\,°C^{-1}$	$\xi_K \times 10^{-6}\,°C^{-1}$	$\xi_{SDS} \times 10^{-6}\,°C^{-1}$
BK7	-0.98	-13.24	-/+12.25
F1	-2.52	-16.44	-/+13.91
Acrylic	-314.41	-128.64	+/-185.78
Polycarbonate	-245.79	-130.04	+/-115.76
NaCl	-92.09	-87.04	+/-5.05
CaF$_2$	-38.3	-36.84	+/-1.46
Si	64.1	-7.44	-/+71.54
Ge	85.19	-11.24	-/+96.42
ZnSe	28.24	-14.44	+/-42.67

The opto-thermal coefficients of diffractive components have negative values, while the OTCs of refractive and SDS components can be either positive or negative.

The nominal optical power of an SDS is equal to zero. Thermally induced changes in SDS optical power are calculated based on **effective optical power** Φ_{eff}^{SDS}, defined by the SDS substrate radius R_0 and the refractive indices of the materials:

$$\Phi_{eff}^{SDS}(\lambda) = -\left[\frac{n_2(\lambda) - n_1(\lambda)}{R_0}\right]$$

The figure shows an example of an athermal hybrid singlet composed of the front aspheric surface shaped to correct for spherical aberrations, and the back SDS surface designed to reduce the singlet dependency on changes in the ambient temperature.

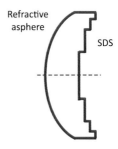

Athermal singlet

Appendix: Diffractive Raytrace

The diffraction of a single ray after encountering a grating surface can be described in vector form as

$$n_2\left(\overline{S'} \times \overline{r}\right) = n_1\left(\overline{S} \times \overline{r}\right) + \Lambda\overline{q} \qquad (A1)$$

where the unit vectors $\overline{S}$, $\overline{S'}$, $\overline{r}$, and $\overline{q}$ are composed of their respective components in Cartesian coordinates:

$$\begin{cases} \overline{S'} = L'_S\overline{i} + M'_S\overline{j} + N'_S\overline{k} \\ \overline{S} = L_S\overline{i} + M_S\overline{j} + N_S\overline{k} \\ \overline{r} = L_r\overline{i} + M_r\overline{j} + N_r\overline{k} \\ \overline{q} = L_q\overline{i} + M_q\overline{j} + N_q\overline{k} \end{cases} \qquad (A2)$$

Unit vectors $\overline{S}$ and $\overline{S'}$ define the ray propagation direction before and after encountering the grating surface; vector $\overline{r}$ defines the local normal to the grating surface; $\overline{q}$ is a vector parallel to the grating grooves at the ray intersection point.

The term $\Lambda\overline{q}$ in Eq. (A1) is responsible for the diffraction phenomenon. For a purely refractive case, the term vanishes, and Eq. (A1) reduces to Snell's law. The **grating parameter** Λ in Eq. (A1) is a function of the working diffraction order m, the **local groove spacing** d_g, and the design wavelength λ_0:

$$\Lambda = m\lambda_0/d_g \qquad (A3)$$

Equation (A1) can be rearranged as

$$\left(n_2\overline{S'} - n_1\overline{S} + \Lambda\overline{p}\right) \times \overline{r} = 0 \qquad (A4)$$

where $\overline{p} = u\overline{i} + v\overline{j} + w\overline{k}$ is a unit vector parallel to the grating substrate and normal to the grating grooves at the ray intersection point $P(x, y)$.

Because the vectors $\overline{q}$ and $\overline{p}$ are orthogonal, we can write

$$\overline{q} = -\overline{p} \times \overline{r} \qquad (A5)$$

The vector S' defining the propagation direction of a ray after diffraction on the grating surface is found from Eq. (A4) in the following form:

$$\overline{S'} = \left(n_1\overline{S} - \Lambda\overline{p} + \Gamma^D\overline{r}\right)/n_2 \qquad (A6)$$

Appendix: Diffractive Raytrace (cont.)

Individual components of the vector $\overline{S}'$ are found from the following equations:

$$\begin{cases} L'_S = \left(n_1 L_S - \Lambda u + \Gamma^D L_r\right) / n_2 \\ M'_S = \left(n_1 M_S - \Lambda v + \Gamma^D M_r\right) / n_2 \\ N'_S = \left(n_1 N_S - \Lambda w + \Gamma^D N_r\right) / n_2 \end{cases} \quad (A7)$$

If $\overline{R} = x\overline{i} + y\overline{j}$ is a radial vector at the intersection point normal to the optical axis, then for any axially symmetric diffractive structure, vectors $\overline{p}$ and $\overline{r}$ are coplanar with vector $\overline{R}$. Therefore, we can write

$$\frac{u}{v} = \frac{L_r}{M_r} = \frac{x}{y} \quad (A8)$$

Components of vector $\overline{p}$ in Cartesian coordinates can be found by

$$\begin{cases} u = \dfrac{xN_r}{\sqrt{x^2+y^2}} \\ v = \dfrac{yN_r}{\sqrt{x^2+y^2}} \\ w = \dfrac{xL_r + yM_r}{\sqrt{x^2+y^2}} = \cos\left(\overline{R},\overline{r}\right) \end{cases} \quad (A9)$$

From Eq. (A9), it follows that $\sin\left(\overline{R},\overline{r}\right) = \pm N_r$. The factor Γ^D in Eqs. (A6) and (A7) depends on the grating type. For the transmission grating, the factor Γ^D is

$$\Gamma^D = -a + \sqrt{(a)^2 - b} \quad (A10)$$

For the reflection grating, the factor Γ^D is

$$\Gamma^D = -a - \sqrt{(a)^2 - b} \quad (A11)$$

Parameters a and b in Eqs. (A10) and (A11) are defined as

$$a = n_1 \cos\left(\overline{r},\overline{S}\right) = n_1 (L_r L_S + M_r M_S + N_r N_S) \quad (A12)$$

$$b = (n_1)^2 - \left[(n_2)^2 + (\Lambda)^2\right] - 2n_1 \Lambda (L_S u + M_S v + N_S w) \quad (A13)$$

In the case of diffractive lenses, the grating spacing $d_g(r)$ is a function of the local radial coordinate r. The local grating

Field Guide to Diffractive Optics

Appendix: Diffractive Raytrace (cont.)

spacing $d_g(r)$ can be used during the analysis of complex multi-element optical systems containing diffractive optics components.

The explicit use of the local grating spacing $d_g(r)$ during diffraction raytrace allows extending the **local grating theory** to optimize performance of an optical system with respect to aberrations, diffraction efficiency, and image contrast.

In the case of diffractive propagation, vectors $\overline{S}$, $\overline{r}$, and $\overline{S'}$ are no longer coplanar, and the diffracted ray does not lie in the **incident plane** defined by the incident ray and the surface normal at the point of intersection. This is fundamentally different from the refraction or reflection propagation phenomenon, where the incident ray, the surface normal at the point of the ray intersection, and the outgoing ray are always coplanar.

The local zone spacing of SDSs $d_{SDS} = d(r, t_0)$ depends on both the radial coordinate r and the step height t_0 and can be found by

$$d_{SDS}(r, t_0) = \frac{\frac{\partial \Phi(r,z)}{\partial r}}{\frac{\partial \Phi(r,z)}{\partial z}} t_0 \tag{A14}$$

The function $\Phi(r,z)$ is the analytical definition of the SDS substrate. When the SDS substrate is explicitly defined as a rotationally symmetric even polynomial aspheric surface with N aspheric terms, vertex curvature c, and conic constant k, the local zone spacing as a function of the radial coordinate can be written as

$$d_{SDS}(r, t_0) = \frac{t_0}{\frac{cr}{\sqrt{1-(1+k)(cr)^2}} + 2\sum_{i=1}^{N} i A_i (r)^{2i-1}} \tag{A15}$$

The local grating parameter Λ, necessary during a diffractive raytrace, is calculated as

$$\Lambda_{SDS}(r, t_0) = \frac{m \lambda_0}{t_0} \left(\frac{cr}{\sqrt{1-(1+k)(cr)^2}} + 2\sum_{i=1}^{N} i A_i (r)^{2i-1} \right)$$

Equation Summary

Diffraction fundamentals:

$$\theta_d \propto \lambda/D \qquad \nabla^2 U(x,y,z) + k_0^2 U(x,y,z) = 0$$

$$U(x_2,y_2,z_2) \propto \iint \frac{1+\cos(\overline{z},\overline{r}_{12})}{2i\lambda r_{12}} \exp(ikr_{12}) U(x_1,y_1,z_1) dx_1 dy_1$$

Fresnel diffraction:

$$U(x_2,y_2,z_2) \propto \frac{\exp(ikz_{12})}{i\lambda z_{12}} \iint \exp\left\{\frac{ik}{2z_{12}}\left[(x_2-x_1)^2 + (y_2-y_1)^2\right]\right\}$$
$$\times U(x_1,y_1,z_1) dx_1 dy_1$$

$$N_F \cong \frac{D^2}{4\lambda L_z} \qquad D = \sqrt{N_F \lambda (4L_z + N_F \lambda)} \qquad L_z = \frac{[D^2 - (N_F \lambda)^2]}{4 N_F \lambda}$$

$$L_T^m = \frac{2md^2}{\lambda} \qquad L_T^{0.5m} = L_T^m - d^2/\lambda = 2(m-0.5)d^2/\lambda$$

Fraunhofer diffraction:

$$U(x_2,y_2,z_2) \propto \frac{\exp(ikz_{12})}{i\lambda z_{12}} \exp\left[\frac{ik\left(x_1^2+y_1^2\right)}{2z_{12}}\right] \times$$
$$\iint \exp\left[\frac{ik(x_2 x_1 + y_2 y_1)}{2z_{12}}\right] U(x_1,y_1,z_1) dx_1 dy_1$$

$$U(x_2,y_2,f) \propto \frac{\exp(ikf)}{i\lambda f} \exp\left[i\frac{k}{2f}\left(x_2^2+y_2^2\right)\right] \times$$
$$\iint \exp\left[-\frac{ik(x_2 x_1 + y_2 y_1)}{2f}\right] U(x_1,y_1,z_1) dx_1 dy_1$$

$$I(r) = I_0 \left(\frac{\pi D^2}{2\lambda f}\right)^2 \left[J_1\left(\frac{\pi D}{\lambda f}r\right)\bigg/\left(\frac{\pi D}{\lambda f}r\right)\right]^2 \qquad D_A \cong 2.44\lambda f/D$$

$$I(q) = I_0 \left[(1-e^2)\frac{qD}{2}\right]^2 \left[\frac{J_1(q)}{q} - e\frac{J_1(eq)}{q}\right]^2$$

$$U(x_2,y_2,f) = (ab)^2 \operatorname{sinc}^2\left(i\frac{kax_2}{2f}\right) \operatorname{sinc}^2\left(i\frac{kay_2}{2f}\right)$$

Volume Bragg gratings:

$$\eta_S = [\sin(\upsilon)]^2 \qquad \eta_P = \{\sin[\upsilon \cos(2\gamma)]\}^2 \qquad \upsilon = \frac{\pi \Delta n T}{\lambda \sqrt{C_R C_S}}$$

$$C_R = \cos(2\gamma) \qquad C_S = \cos(2\gamma) - \frac{\lambda}{n d_B}\tan\left(\frac{\beta-\alpha_B}{2}\right)$$

Field Guide to Diffractive Optics

Equation Summary

$$\Delta n = \frac{\lambda(s - 0.5)}{T}\sqrt{\cos(2\gamma)\left[\cos(2\gamma) - \frac{\lambda}{nd_B}\tan\left(\frac{\beta - \alpha_B}{2}\right)\right]}$$

$$\beta = \cos^{-1}\left(\frac{2p-1}{2s-1}\right) - \alpha_B \qquad \beta = 180 - \cos^{-1}\left(\frac{2p-1}{2s-1}\right) - \alpha_B$$

Grating equation:

$$n_2 \sin\theta_m = n_1 \sin\theta_i + m\lambda/d_g \qquad \sin\theta_m + \sin\theta_i = m\lambda/d_g$$

$$\sin\theta_m^L = m\lambda/2d_g$$

Grating properties:

$$\frac{\partial\theta_m}{\partial\lambda} = \frac{m}{d_g n_2 \cos\theta_m} \qquad \frac{\partial\theta_m^L}{\partial\lambda} = \frac{m}{2d_g \cos\theta_m^L} = \frac{\tan(\varphi_b)}{\lambda_b^L}$$

$$\frac{\partial\theta_m}{\partial\lambda} = \frac{n_2 \sin\theta_m - n_1 \sin\theta_i}{\lambda n_2 \cos\theta_m} = \frac{1}{\left[\frac{d_g n_1 \cos(\theta_i)}{m} - \frac{\lambda_b}{\tan(\varphi_b)}\right]}$$

Free spectral range:

$$\Delta\lambda_{FSR} = \frac{\lambda_s}{m} = \frac{\lambda_l}{m+1} \qquad \frac{\lambda}{d\lambda} = |m|N_g \qquad \left(\frac{\lambda}{d\lambda}\right)_{max} = \frac{2W_g}{\lambda}$$

$$\frac{\lambda}{d\lambda} = |m|\frac{W_g}{d_g} = \frac{W_g}{\lambda}|\sin\theta_m + \sin\theta_i|$$

Grating anomalies:

$$\left|\frac{1}{n_2}\left(n_1 \sin\theta_i + \frac{m\lambda}{d_g}\right)\right| = 1 \qquad \lambda = \frac{d}{m}[\text{sgn}(m) - \sin(\theta_i)]$$

Blazing condition:

$$\begin{cases} n_1 \sin(\theta_i + \varphi) = n_2 \sin(\theta_m + \varphi_b) \\ n_2 \sin(\theta_m) = n_1 \sin(\theta_i) + m\lambda/d_g \end{cases}$$

$$\lambda_b = \frac{d_g}{m}\left(n_2 \sin\left\{\sin^{-1}\left[\frac{n_2}{n_1}\sin(\theta_i + \varphi_b)\right] - \varphi_b\right\} - n_1 \sin(\theta_i)\right)$$

$$\lambda_b = \frac{2d_g}{m}\sin(\varphi_b)\sin(\theta_i + \varphi_b) \qquad \lambda_b^L = \frac{2d_g}{m}\sin(\varphi_b)$$

Field Guide to Diffractive Optics

Equation Summary

Blazed facet angle and height:

$$\tan(\varphi_b) = \frac{(m\lambda_b/d_g)}{n_1\cos(\theta_i) - \sqrt{(n_2)^2 - [n_1\sin(\theta_i) + (m\lambda_b/d_g)]^2}}$$

$$\phi = d_g[n_2\sin(\theta_m) - n_1\sin(\theta_i)] = m\lambda_b$$

$$h_{opt} = \left| m\lambda_b \middle/ \left\{ \sqrt{(n_2)^2 - [n_1\sin(\theta_i) + (m\lambda_b/d_g)]^2} - n_1\cos(\theta_i) \right\} \right|$$

$$h_{opt} = \left| m\lambda_b \middle/ \left[\sqrt{(n_2)^2 - (m\lambda_b/d_g)^2} - n_1 \right] \right|$$

$$\varphi_p = \cos^{-1}[n_1\sin(\theta_i)/n_2 + m\lambda_b/(n_2 d_g)]$$

Scalar grating theory:

$$\eta_m = \left| \frac{1}{d_g} \int_0^{d_g} t(x) e^{i\frac{2\pi}{\lambda}\left\{[n_1\cos(\theta_i) - n_2\cos(\theta_m)]f(x) - \frac{m\lambda}{d_g}x\right\}} dx \right|^2$$

$$\eta_m = \left\{ \sin\left[m\pi\left(\frac{\lambda_b}{\lambda} - 1\right)\right] \middle/ \left[m\pi\left(\frac{\lambda_b}{\lambda} - 1\right)\right] \right\}^2$$

$$\eta_M = HI_S = \left[\sin\left(M\frac{kdp}{2}\right) \middle/ M\sin\left(\frac{kdp}{2}\right)\right]^2 \left[\sin\left(\frac{kdp}{2}\right) \middle/ \frac{kdp}{2}\right]^2$$

$$\eta = \left\{ \sin\left[m\pi\left(\frac{\lambda_b}{\lambda} - 1\right)\right] \middle/ m\pi\left(\frac{\lambda_b}{\lambda} - 1\right) \right\}^2 \left[\sin\left(\frac{\pi\lambda_b}{\lambda N}\right) \middle/ \left(\frac{\pi\lambda_b}{\lambda N}\right)\right]^2$$

Extended scalar theory:

$$\eta_{FF} \approx \zeta\eta_{scalar} \qquad \zeta = \frac{d_m}{d_g} = 1 - \frac{[n_1\sin(\theta_i) + m\lambda/d_g]\tan(\varphi)}{\sqrt{(n_2)^2 - (n_1\sin\theta_i + m\lambda/d_g)^2}}$$

$$\zeta = 1 - \frac{1}{\sqrt{(n_2)^2 - (n_1\sin\theta_i + m\lambda/d_g)^2}}$$
$$\times \frac{[n_1\sin(\theta_i) + m\lambda/d_g](m\lambda/d_g)}{\left(\left|n_1\cos(\theta_i) - \sqrt{(n_2)^2 - [n_1\sin(\theta_i) + m\lambda/d_g]^2}\right|\right)}$$

$$\zeta = 1 - \frac{(m\lambda/d_g)^2}{\sqrt{(n_2)^2 - (m\lambda/d_g)^2}\left(\left|n_1 - \sqrt{(n_2)^2 - (m\lambda/d_g)^2}\right|\right)}$$

Equation Summary

Rigorous analysis:

$$\eta_{rig} \approx \eta_{scalar}\left[1 - C_m(n,\theta)\frac{\lambda_B}{d}\right]$$

$$\frac{d}{\lambda} < \frac{1}{\sqrt{\max(n_1, n_2) + n_1 \sin(\theta_i)}}$$

Effective medium theory:

$$n_\perp = \frac{n_1 n_1}{\sqrt{(n_1)^2 \zeta_g + (n_2)^2 (1 - \zeta_g)}}$$

$$n_\parallel = \sqrt{(n_1)^2 \zeta_g + (n_2)^2 (1 - \zeta_g)}$$

Grating doublets:

$$OPD = m\lambda_b = h[n_2(\lambda)\cos(\theta_m) - n_1(\lambda)\cos(\theta_i)]$$

$$\frac{d}{d\lambda}[n_2(\lambda)]\cos(\theta_m) - \frac{d}{d\lambda}[n_1(\lambda)]\cos(\theta_i) = m$$

$$OPD = m\lambda_{bl} = h_1[n_3(\lambda)\cos(\theta_{m1}) - n_1(\lambda)\cos(\theta_{i1})] + h_2[n_2(\lambda)\cos(\theta_{m2}) - n_3(\lambda)\cos(\theta_{i2})]$$

$$h_1\left\{\frac{d}{d\lambda}[n_3(\lambda)]\cos(\theta_{m1}) - \frac{d}{d\lambda}[n_1(\lambda)]\cos(\theta_{i1})\right\} +$$

$$h_2\left\{\frac{d}{d\lambda}[n_2(\lambda)]\cos(\theta_{m2}) - \frac{d}{d\lambda}[n_3(\lambda)]\cos(\theta_{i2})\right\} = m$$

$$\frac{d}{d\lambda}[n_2(\lambda)]h_2\cos(\theta_{m2}) - \frac{d}{d\lambda}[n_1(\lambda)]h_1\cos(\theta_{i1}) = m$$

$$\Delta OPD = \Delta h_1[n_1(\lambda) - n_2(\lambda)] \qquad \Delta OPD = \Delta h_1[n_1(\lambda) - 1]$$

Diffractive lens surfaces:

$$\Phi^D(\lambda) = 8N_k\lambda_0/(D_0)^2 \qquad \phi = m\lambda_0 \qquad \Phi^D(\lambda) = \Phi_0^D \frac{\lambda}{\lambda_0}$$

$$h(\theta_i) = \frac{\lambda_0}{\sqrt{(n_0)^2 - (\sin\theta_i)^2} - \cos\theta_i}$$

$$t(r, \theta_i) = h(\theta_i)[\phi(r) \bmod 2\pi] \qquad h_0 = \lambda_0/(n_0 - 1)$$

$$t_b = \frac{h_0}{2^N} = \frac{\lambda_0}{(n_0 - 1)2^N} \qquad \Delta f_{chr}^D = \Delta\lambda/\left(\Phi_0^D \lambda_0\right)$$

Field Guide to Diffractive Optics

Equation Summary

$$\psi(x,y) = \frac{2\pi}{\lambda_0} \left\{ \left[\sqrt{(x-x_o)^2 + (y-y_o)^2 + (z_o)^2} - z_o \right] - \left[\sqrt{(x-x_i)^2 + (y-y_i)^2 + (z_i)^2} - z_i \right] \right\}$$

$$\psi(x,y) = m \sum_{k=0}^{K} \sum_{l=0}^{L} A_{kl}(x)^k (y)^l \qquad \psi(r) = m \sum_{i=0}^{N} A_i (r)^{2i}$$

Stepped diffractive surfaces:

$$\Phi^{SDS}(\lambda) = \left| \frac{n_2(\lambda) - n_1(\lambda)}{R_0} \right| \left(\frac{\lambda}{\lambda_0} - 1 \right)$$

$$h_{SDS} = \left| \frac{m\lambda_0}{\sqrt{(n_2)^2 - [\sin(\theta_i)]^2} - n_1 \cos(\theta_i)} \right| \qquad h_{SDS} = \left| \frac{m\lambda_0}{n_2 - n_1} \right|$$

Multi-order diffractive lenses:

$$h_m = \frac{m\lambda_0}{n(\lambda_0) - 1} \qquad \frac{m\lambda_B}{n(\lambda_B) - 1} = \frac{k\lambda_R}{n(\lambda_R) - 1}$$

Hybrid diffractive lenses:

$$\Phi^H(\lambda) = \Phi^R(\lambda) + \Phi^D(\lambda) = \Phi_0^R \left[1 + \frac{D_n(\lambda)\Delta\lambda}{n_0(\lambda) - 1} \right] + \Phi_0^D \left(1 + \frac{\Delta\lambda}{\lambda_0} \right)$$

$$\frac{\Phi_0^R}{\Phi_0^D} = \frac{\Delta\lambda[n_0(\lambda) - 1]}{\lambda_0[n_0(\lambda) - n(\lambda)]} = -\frac{[n_0(\lambda) - 1]}{\lambda_0 D_n(\lambda)}$$

$$\Phi^{AH} = \Phi_0^D \left\{ 1 - \frac{[n_0(\lambda) - 1]}{\lambda_0 D_n(\lambda)} \right\} = \Phi_0^R \left\{ 1 - \frac{\lambda_0 D_n(\lambda)}{[n_0(\lambda) - 1]} \right\}$$

Opto-thermal properties:

$$\xi_K = -2\alpha \qquad \xi_{SDS}(\lambda) = \pm[\xi_R(\lambda) - \xi_K]$$

$$\xi_R(\lambda) = \frac{1}{[n_2(\lambda) - n_1(\lambda)]} \left[\frac{dn_2(\lambda)}{dT} - \frac{dn_1(\lambda)}{dT} \right] - \alpha$$

$$\xi_{SDS}(\lambda) = \frac{1}{[n_2(\lambda) - n_1(\lambda)]} \left[\frac{dn_2(\lambda)}{dT} - \frac{dn_1(\lambda)}{dT} \right] + \alpha$$

Field Guide to Diffractive Optics

Equation Summary

Athermalization with diffractive components:

$$\frac{\Delta \Phi}{\Phi} = \xi \Delta T \qquad \frac{\Phi^K}{\Phi^R} = \frac{\left\{\frac{d}{dT}[n_2(\lambda)] - \frac{d}{dT}[n_1(\lambda)]\right\}}{2\alpha[n_2(\lambda) - n_1(\lambda)]} - \frac{1}{2}$$

$$\Phi_{eff}^{SDS}(\lambda) = -\left[\frac{n_2(\lambda) - n_1(\lambda)}{R_0}\right]$$

Bibliography

Arieli, Y. et al., "Design of a diffractive optical element for wide spectral bandwidth," *Opt. Lett.* **23**(11), 823–825 (1998).

Ersoy, O. K., *Diffraction, Fourier Optics and Imaging*, John Wiley & Sons, Inc., Hoboken, NJ (2007).

Farn, M. W. and W. B. Veldkamp, "Binary optics," in *Handbook of Optics*, Vol. II, M. Bass, Ed., Ch. 8, McGraw-Hill, New York (1995).

Greisukh, G. I., S. T. Bobrov, and S. A. Stepanov, *Optics of Diffractive and Gradient-Index Elements and Systems*, SPIE Press, Bellingham, WA (1997).

Hutley, M. C., *Diffraction Gratings and Applications*, Academic Press, London (1982).

Ichikawa, H., K. Masuda, and T. Ueda, "Analysis of micro-Fresnel lenses with local grating theory and its comparison with fully electromagnetic methods," *J. Opt. Soc. Amer. A* **26**, 1938–1944 (2009).

Joannopoulos, J. D., S. G. Johnson, J. N. Winn, and R. D. Meade, *Photonic Crystals. Molding the Flow of Light*, 2nd ed., Princeton University Press, Princeton, NJ (2008).

Jordan, J. A., P. M. Hirsch, L. B. Lesem, and D. V. Van Rooy, "Kinoform lenses," *Appl. Opt.* **9**, 1883–1887 (1970).

Koronkevich, V. P., "Computer synthesis of diffraction optical elements," in *Optical Processing and Computing*, H. H. Arsenault, T. Szoplik, and B. Macukow, Eds., Academic Press, San Diego, 277–313 (1989).

Kogelnik, H., "Coupled wave theory for thick hologram gratings," *Bell Syst. Tech. J.* **48**, 2909–2948 (1969).

Kress, B. and P. Meyrueis, *Digital Diffractive Optics*, John Wiley & Sons, Inc., Hoboken, NJ (2000).

Lalanne, P., S. Astilean, P. Chavel, E. Cambril, and H. Launois, "Design and fabrication of blazed binary diffractive elements with sampling periods smaller than the structural cutoff," *J. Opt. Soc. Amer. A* **16**, 1143–1156 (1999).

Bibliography

Lee, W.-H., "Computer-generated holograms: techniques and applications," in *Progress in Optics XVI*, E. Wolf, Ed., North-Holland, Amsterdam, 119–232 (1978).

Lesem, L. B., P. M. Hirsch, and J. A. Jordan, "The kinoform: A new reconstruction device," *IBM J. Res. Dev.* **13**, 150–155 (1969).

Lipson, S. G., H. Lipson, and D. S. Tannhauser, *Optical Physics*, Cambridge University Press, Cambridge, UK (2001).

Loewen, E. G. and E. Popov, *Diffraction Gratings and Applications*, Marcel Dekker, Inc., New York (1997).

Novotny, L. and B. Hecht, *Principles of Nano-Optics*, Cambridge University Press, Cambridge, UK (2006).

Palmer, C. A. and E. G. Loewen, *Diffraction Grating Handbook*, David Richardson Grating Laboratory, Rochester, NY (2000).

Pommet, D. A., M. G. Moharam, and E. B. Grann, "Limits of scalar diffraction theory for diffractive phase elements," *J. Opt. Soc. Amer. A* **11**, 1827–1834 (1994).

Raguin, D. H., S. Norton, and G. M. Morris, "Subwavelength structured surfaces and their applications," in *Diffractive and Miniaturized Optics*, S. H. Lee, Ed., SPIE Critical Reviews **CR49**, SPIE Press, Bellingham, WA, 234–265 (1993).

Soskind, Y. G., "Novel technique for passive athermalization of optical systems," *OSA Trends in Optics and Photonics, Diffractive Optics and Micro-Optics* **43**, 194–204 (2000).

Spencer, G. H. and M. V. R. K. Murty, "Generalized ray-tracing procedure," *J. Opt. Soc. Amer.* **41**, 672–678 (1951).

Stone, T. and N. George, "Hybrid diffractive-refractive lenses and achromats," *Appl. Opt.* **27**, 2960–2971 (1988).

Swanson, G. L., "Binary optics technology: The theory and design of multi-level diffractive optical elements," *Lincoln Laboratory Tech. Report* **854** (1989).

Swanson, G. L., "Binary optics technology: Theoretical limits on the diffraction efficiency of multilevel diffractive optical elements," *Lincoln Laboratory Tech. Report* **914** (1991).

Bibliography

Traub, W. A., "Constant-dispersion grism spectrometer for channeled spectra," *J. Opt. Soc. Amer.* **7**(9), 1779–1791 (1990).

Veldcamp, W. B., G. J. Swanson, and D. C. Shaver, "High efficiency binary lenses," *Opt. Comm.* **53**, 353–358 (1985).

Wyrowski, F. and O. Bryngdahl, "Digital holography as part of diffractive optics," *Reports on Progress in Physics* **54**, 1481–1571 (1991).

Index

achromatic condition, 98
achromatic hybrid structure, 98
achromatic refractive doublet, 98
air-spaced grating doublet, 79
Airy beams, 31
Airy disk, 14, 29
Airy distribution, 7
Airy pattern, 13, 14
amplitude division, 37
amplitude filters, 33
amplitude gratings, 36
amplitude masks, 29, 87
angular dispersion, 49
angular switching, 53
annular phase plates, 87
aperiodically spaced aperture (ASA), 27
aperture spacing, 25
apertures with central obscuration, 20
apodization, 19
array beam generators, 43
artificial dielectrics, 41, 67, 69
athermal achromat, 100
athermal condition, 100
athermal lens, 100
athermalization, 94
autocollimation, 48, 49
axial chromatic aberration, 91

band diagram, 47
band-pass filter, 54
beam obscuration, 15
bidirectional propagation, 87

binary diffractive lenses, 90
binary level, 88
binary phase gratings, 44, 53, 62
binary surface structures, 69, 88
blazed binary grating, 67
blazed facet angle, 56, 57, 88
blazed gratings, 56
blazed transmission gratings, 71
blazing condition, 40, 49, 56
blazing wavelength, 56
Bloch states, 47
Bragg phenomenon, 47
Bragg plane, 37
Brillouin zones, 47
broadband blazing, 77, 78
broadband diffraction efficiency, 83, 84

central core, 15
central obscuration, 5, 16
classical mount, 48
coefficient of thermal expansion (CTE), 99
coherent illumination, 28
complex amplitude, 2
computer-generated hologram, 46
conical mount, 48
contrast reduction, 17
convolution, 28
coupled wave analysis, 70
cut-off orders, 51

Dammann grating, 44, 53
design wavelength, 89, 94
diffraction, 1

Index

diffraction efficiency, 10, 61
diffraction gratings, 3, 36
diffraction-limited (lens performance), 13
diffraction rings, 32
diffractive beam-shaping components, 45
diffractive doublet, 77
diffractive homogenizers, 42
diffractive kinoforms, 88, 89
diffractive lens doublet, 96
diffractive lens surface (DLS), 87, 88
diffractive optical power, 81
diffractive phase cells, 43
diffractive polarizer, 55
diffractive singlet, 77, 78
digital diffractive optical element, 45, 46
doughnut-shaped field, 35
duty cycle, 26, 63, 64

echelle, 95
effective medium theory, 67, 69
effective optical power, 101
effective refractive index, 41
efficiency angular shift, 76
elliptical distribution, 17
etalon effect, 54
etching, 89
evanescent orders, 51
extended depth of field, 31
extended object, 28
extended scalar diffraction theory, 63, 65

facet angle, 39
facet width, 84

fan-out elements, 43
far field, 3, 14
far-field shaping components, 42, 43, 45
fill factor, 26, 63
finite difference method, 70
finite difference time domain (FDTD), 41, 70
fluorescence depiction microscopy, 34
focus, 1
form birefringence, 69, 73
Fourier grating, 43
Fourier transform lens, 45
fractional Talbot distributions, 11, 12
Fraunhofer approximation, 3
Fraunhofer diffraction, 3, 14
free spectral range, 50
frequency comb, 86
Fresnel approximation, 3
Fresnel diffraction integral, 3
Fresnel lens, 89
Fresnel phase plate (FPP), 8
Fresnel reflections, 63, 70
Fresnel zone number, 4
Fresnel zone plate (FZP), 6
Fresnel zones, 4

Gaussian lens formula, 92
ghost orders, 64
grating doublet, 77, 78
grating equation, 48
grating parameter, 102
grating resolution, 50
grayscale apodizer, 23
grazing incidence, 49

Field Guide to Diffractive Optics

Index

grazing mount, 49
GRISM, 40
groove spacing, 39

Helmholtz equation, 2
holographic diffuser, 42
holographic optical element, 46
holographic recording, 42
Huygens-Fresnel principle, 2, 3
hybrid achromat, 98
hybrid components, 97
hybrid structures, 88, 97

ideal lens, 9
incoherent illumination, 28
incident plane, 52, 104
integral method, 70
interference, 37
isofrequency diagram, 47

kinoform, 43, 88
Kirchhoff's diffraction integral, 3

Lambertian scatter, 42
lamellar grating, 39
lateral chromatic aberration, 91
lattice constant, 47
lens transfer function, 13
linear dispersion, 49
linear gratings, 36
lithographic techniques, 62
Littrow mounting, 48, 56
local grating theory, 104
local groove spacing, 102
longitudinal chromatic aberration, 91

material dispersion, 91
Maxwell's equations, 2, 66
modal analysis, 70
monolithic grating doublet, 80
multi-order diffractve lenses, 95
multiple apertures, 25
multispot beam generator, 43

near-field shaping components, 45
negative refraction, 47

observation point, 4
opaque semiplane, 21
optical path difference (OPD), 30, 58, 78, 80
optical power, 88, 89, 91, 93, 94
optical tweezers, 34
optimum profile height, 58, 59
optimum zone height, 89
opto-thermal coefficient (OTC), 99

P-polarized (light), 37, 38, 52
paraxial approximation, 3
passing-off orders, 51
passive facet angle, 56, 59
pattern distortion, 17
peak diffraction efficiency, 73
phase delay, 88
phase filters, 33
phase gratings, 36
phase mask, 28, 29
phase profile, 92

Field Guide to Diffractive Optics

Index

photonic bandgap, 47
photonic crystal, 41, 47
photoresist, 39
point spread function (PSF), 13
point spread function engineering, 31
polarization anisotropy, 52, 75
polarization dependency, 52, 72, 84
polarization extinction ratio, 55
polychromatic diffraction efficiency, 77, 81, 85
polychromatic efficiency modulation, 85
polychromatic point spread function, 81
propagating orders, 51
pupil filter, 29, 33
pupil mask, 29

Rayleigh anomalies, 52
Rayleigh resolution criterion, 28
rectangular aperture, 18
reflection, 36, 87
refractive achromat, 98
refractive doublet, 81
refractive optical power, 81
relative diffraction efficiency, 61
relative feature size, 68
resolution, 1, 15
resonant domain, 67, 68
rigorous diffraction analysis, 65, 66, 70

S-polarized (light), 37, 38, 52

sagittal plane, 52
scalar diffraction theory, 2, 60, 70
scalar domain, 60, 73
serrated aperture, 23
serrated edge, 22
shadowing effect, 63, 93
single-point diamond turning (SPDT), 93
sinusoidal grating, 39
Snell's Law, 48
soft-edge aperture, 19
soft-edge apodizer, 22
spaced grating doublet, 79, 82
Sparrow resolution criterion, 28
speckle, 45
spectral bandwidth, 91
spectral shift, 74
spurious diffraction orders, 96
staircase lenses, 93
star test, 13
stepped diffractive surface (SDS), 87, 93
Strehl ratio, 29, 30
structural cutoff, 67
supercollimation, 47
superprism effect, 47
superresolved PSF, 29, 35
surface relief, 36

Talbot distance, 11
Talbot effect, 11
Talbot image, 11
Talbot plane, 11
tangential plane, 52
TE-polarized (light), 37, 52

Field Guide to Diffractive Optics

Index

thin-element approximation, 60
three-dimensional diffractive structures, 47
threshold condition, 51
TM-polarized (light), 37, 52
topological charge, 34
Toraldo di Francia, 29
transfer etching, 39
transmission, 36, 87
transverse chromatic aberration, 91
triangular grooves, 39
two-dimensional diffractive structures, 41

very large-scale integration (VLSI), 90
volume Bragg gratings (VBG), 37, 47
volumetric anisotropy, 47
vortex phase masks, 31, 34

wave equation, 60
wave vector diagrams, 47
wavenumber, 2
Wood anomalies, 51

zero deflection, 40
zero-order gratings, 69

Yakov G. Soskind is Principal Systems Engineer for DHPC Technologies, Inc., in Woodbridge, NJ. He provides innovative solutions and technical expertise to customers in the areas of laser optics, electro-optical sensors, photonics instrumentation, and system-level integration.

Dr. Soskind is a recognized expert in the fields of optical design, diffractive optics, laser systems, and illumination. For more than 30 years, he has contributed to the fields of electro-optical and photonics engineering, successfully reducing to practice numerous innovative solutions in the form of fiber optics and photonics devices, diffractive structures, laser optics, imaging, and illumination devices. His innovative work in the field of diffractive optics has led to the development of novel types of diffractive structures that have enhanced diffraction efficiency and are covered in several issued patents.

Dr. Soskind has been awarded more than 20 domestic and international patents and has authored and co-authored several publications and conference presentations. He also serves on the technical committees of two international conferences.